The Castle –

Implementing ERP through business change

by

Jim Hourigan

and

Gary Jenkins

All rights reserved. No part of this work may be reproduced or stored in an information retrieval system (other than for purposes of review) without the express permission of the authors in writing.

First published 2005 British Library Cataloguing in Publication Data A catalogue record for this book is available from the British Library

Note: The material contained in this book is set out in good faith for general guidance and no liability can be accepted for loss or expense incurred as a result of relying in particular circumstances on statements made in the book.

Names, characters, places and incidents are either the product of the authors' imaginations or are used fictitiously. Any resemblance to actual events or places and people, living or dead, is entirely coincidental.

© Copyright 2005 Jim Hourigan and Gary Jenkins.

All rights reserved. No part of this publication may be reproduced, stored in a retrieval system, or transmitted, in any form or by any means, electronic, mechanical, photocopying, recording, or otherwise, without the written prior permission of the author.

Note for Librarians: a cataloguing record for this book that includes Dewey Decimal Classification and US Library of Congress numbers is available from the Library and Archives of Canada. The complete cataloguing record can be obtained from their online database at:
www.collectionscanada.ca/amicus/index-e.html
ISBN 1-4120-2656-3

TRAFFORD *Offices in Canada, USA, Ireland and UK*

This book was published *on-demand* in cooperation with Trafford Publishing. On-demand publishing is a unique process and service of making a book available for retail sale to the public taking advantage of on-demand manufacturing and Internet marketing. On-demand publishing includes promotions, retail sales, manufacturing, order fulfilment, accounting and collecting royalties on behalf of the author.

Book sales for North America and international:
Trafford Publishing, 6E–2333 Government St.,
Victoria, BC v8t 4p4 CANADA
phone 250 383 6864 (toll-free 1 888 232 4444)
fax 250 383 6804; email to orders@trafford.com

Book sales in Europe:
Trafford Publishing (uk) Ltd., Enterprise House, Wistaston Road Business Centre,
Wistaston Road, Crewe, Cheshire cw2 7rp UNITED KINGDOM
phone 01270 251 396 (local rate 0845 230 9601)
facsimile 01270 254 983; orders.uk@trafford.com

Order online at:
trafford.com/04-0484

10   9   8   7   6   5   4   3   2

Contents

Prologue

Acknowledgements

Foreword

Chapter 1	The Need for Change
Chapter 2	How to start
Chapter 3	Define the boundaries
Chapter 4	How will the project work?
Chapter 5	First view of objectives and deliverables
Chapter 6	Where are we now?
Chapter 7	Set strategy and vision
Chapter 8	How to help the organisation change – mobilisation and kick-off
Chapter 9	Gathering the Data
Chapter 10	Establishing the 'as is'
Chapter 11	Processes
Chapter 12	Detailed Design (Part 1)
Chapter 13	Preparation
Chapter 14	Problems and Input
Chapter 15	Costs and Benefits
Chapter 16	Identify main impacts of change
Chapter 17	Agree scale and phasing of changes
Chapter 18	Detailed Design Part 2
Chapter 19	Implementation
Chapter 20	Changes
Chapter 21	Continuous improvement
Chapter 22	Lessons learned
Chapter 23	Realising the Benefits

Acknowledgements

We've written this book with lots of help from Bernice Walmsley and would like to thank her for her work, her input and, above all, her patience. Apart from her assistance with the writing, she has also kindly supplied the drawing of the castle on the front of our book.

We'd also like to thank APICS – the Educational Society for Resource Management (American Production and Inventory Control Society) for their help with defining terms and their kind permission to reprint their materials.

Thanks are due to Elitehelp Limited and Columbia Crest Consulting for all their input, practical knowledge and helpful comments. Without their consulting experience this book would have been impossible to write.

We are also indebted to all the organisations and people with whom we've worked over the years for each and every experience that we've had – both good and bad – as all experience is useful and character forming. Many of these experiences have found their way into this book.

Jim Hourigan and Gary Jenkins 2004

Foreword

'A castle – a fortification and a home'
                                    Anon.

We all know that change is difficult and can be very painful. But there is no escaping it; at some time in the near future your organisation will face the daunting task of implementing or upgrading your ERP systems. When this time comes it is vitally important that everyone understands the process and its implications.

So, how can this book help you to understand and – just as importantly – to pass on this understanding to everyone in your organisation? If you miss the opportunity to educate people about ERP, you will be like the young boy with his finger in the dam; one leak and the dam weakens, two or more leaks lead to greater weakness and soon the whole dam is in question. How will you ever get everyone to understand?

ERP implementation is a wide and complicated subject – it needs simplification and explanation. An ERP system is so much more than just software. Software is an enabler but an ERP system is how your organisation will structure its processes for effective planning and control. In essence, with ERP, you are creating a castle – a fortification against your opponents (your competition) and you are building a home – a place of security, familiarity and constant change providing all who live within its walls an environment that encourages support, challenges without fear and demands growth.

Knowledge and experience will help but 'The Castle' will provide invaluable support. In an easy-going narrative 'The Castle' juxtaposes the trials and triumphs of an ERP system implementation with the family stresses of the building of a new home.

The objective of this book is to provide a story – and a method for change - that most people can relate to. The story is about a newly merged family building a home. That story, integrated with an account of a newly merged company implementing an ERP system, will provide you with the analogies to recognise the issues your organisation will face. These analogies provide an easy read while giving you an understanding of how the principles of change management will help you to help your organisation and also the realisation of just how comprehensive an ERP implementation can be.

The authors have seen the good, the bad and the truly ugly of implementing and upgrading ERP systems. The goal of 'The Castle' is to help you to gain a greater understanding of what you will be facing when undertaking this change – and to use this understanding to help avoid some of the most common problems with an ERP system. There will be problems but the aim is to minimise them.

For those of you fortunate enough to be chosen, drafted or volunteered to lead an ERP program, recognise that you will need abundant patience and that even people who love you will hate you at times but that there will also be moments – usually fleeting – of genuine, astonishing clarity and unbelievable luck.

Chapter 1 – The Need for Change

'For goodness sake! How much longer are you going to be? Come out, now!' yelled Jim, shuffling from foot to foot. 'Please.' he added, as an afterthought. After all, they had only been living together for a few days.
The problems had become apparent immediately. There just wasn't the room in this three-bedroom town house for two families. He had tried to be patient, he really had, but this first day back at work after his honeymoon was already proving just too much for him.
Jan rushed by him in her new dressing gown and she gave him a squeeze as she edged past him on the narrow landing.
'Eggs and bacon or just cereal?' she asked as she swung round the banister and went down the stairs with even more than her usual speed.
'I'll take a banana with me, thanks.' he answered. 'There'll be no time for eating when I've finally had my turn in the bathroom.'
'Who's in there anyway? Coffee's on.' She said as she swept back to their bedroom two minutes later. He was still waiting, with increasing impatience, outside the bathroom while his newly acquired stepdaughter took her time in making herself beautiful for school. Surely fourteen year olds didn't usually take this long to get ready, he thought. He supposed that make-up was not allowed at school so what else did she need to do apart from have a quick shower?
'It's Emma' he said ' she's been in there for about fifteen minutes. I'm going to be late.'
He gave up and dashed back into the bedroom where he retrieved papers from the little corner desk under the eaves and stuffed them into his briefcase. As he grasped another handful, he glanced at them but couldn't really remember what he'd been doing with them before the wedding two weeks ago. He heard the

bathroom door creak shut and he put his briefcase down.
He dashed back and tried to open the door but it was locked again.
'Emma. Come out this minute'
'It's not me' said a muffled voice from the girls' bedroom
'Well, who is it then?'
'Not me' chorused two other voices.
'It's me' said Philippa as she opened the bathroom door and swept past her father.
Jim was getting more and more grumpy by the minute. Finally, he made it into the bathroom. He looked at himself in the mirror. His hair was standing on end as though he'd just got out of bed, but other than that, he looked wide awake. No wonder, I've been up for hours standing on the landing, he thought. He zipped through his usual routine in five minutes flat, dressed, grabbed a banana and half a cup of coffee on his way through the kitchen and raced out of the door, leaving his new wife to sort out the families – family, singular, he corrected himself – as he got into his black BMW and swung it into the cul-de-sac.
Arriving at work, he was greeted with wary looks from his colleagues. That's a sure sign that things haven't gone well while I've been away, he thought. He hadn't even unpacked his briefcase or switched on his computer to check his e-mails, before his boss poked his head round the door and asked him to meet him in the boardroom at 9.30.
'What's up?' said Jim as he sat down opposite his boss at the highly polished table.
'You haven't heard then?' said Neil, rubbing his chin and obviously wondering where to start.
Jim began to feel distinctly uneasy. This was obviously not good news. He'd worked at MMB for ten years now. Since just before his first marriage broke up.

'We're going to merge with Abbey Products,' said Neil Martin, the Managing Director of Martin Milling 'here's the Press Release going out today. All the other staff were told on Friday but, obviously, you weren't here so this is the first opportunity we've had to talk. Anyway, did everything go well?'

The honeymoon and the wedding seemed a million years ago now and Jim mumbled a distracted reply as he tried to digest the news and figure out what it might mean.

'Does that mean jobs will go?' he looked straight at Neil to gauge his reply but knew from his long experience of working for the man that he would get an honest answer.

'Well, some obviously, but we've a long way to go before we get to that point. What do you think?'

'Hell, I knew that you were looking at a few changes, but this! How do you feel about it all?'

'It can only be good news for both companies. We've both got a good deal. Anthony Abbey is a shrewd operator. He's built a good business and it fits well with us. They're really strong on export – that's where we'll gain. There'll be some difficult times ahead but it's a good move for us. You'll be OK Jim.'

Jim sighed with relief but thought to himself 'Well, he would say that wouldn't he?' He pulled himself together; aware that now was not the time to let his mind wander. Neil was giving him more news.

'What I want you to do is head up the Merger Committee at the MML end of things. There'll be you from Purchasing, Bill from Finance, Ron from Logistics, Malcolm from IT, Gerry from Sales, Katrina from Marketing, David from Production and Caroline from HR. I don't know yet all the details about who Abbey will be nominating. It will mean going through everything to see where we need to change – all the systems, each department one by one to see what needs doing to create one company.'

Jim's mind was racing and he realised that a fine film of sweat was rapidly cooling his forehead. Was this a new job, was it a good move, could he handle it? But it occurred to him that this did not sound quite right. 'All the systems? Shouldn't Malcolm be heading this up then? I'm no expert on software, you know.'

'I realise that but I didn't mean just the software. From what I understand from Tony Abbey, it's not just the software or the staff that need sorting out, it's the whole thing – the way the two companies work, the processes and so on. It's a big job but I think you're the man for the job here. You won't be on your own though. The Merger Committee will be looked after overall by Tony Abbey – he'll be acting as Programme Sponsor and he'll be reinforced by John Forsythe and a whole team from Abbey who will also be on the Merger Committee.'

'Well, great. I'm pleased you think I can handle it. Where do we start?' Both men looked up as the door opened and two men entered.

'Oh, hello John. Thanks for coming along. You know Jim, my Purchasing Manager don't you?' The tall, grey-haired man who had entered the office without knocking extended his hand to Jim. They had met before, briefly, Jim remembered, when John Forsythe had visited Martin's some months previously. He hadn't been told on that occasion what John's business was.

'John is Abbey's Finance Director and he'll be heading up the work on the merger at the Abbey side of things so you'll need to work closely together.' Neil turned to the other man – a younger man with tightly curled blond hair who seemed to have a natural air of authority and energy despite the fact that he hung back, waiting to be introduced. 'And this is Tony Abbey.'

Tony and John joined the two Martin men at the highly polished table in Neil's large, immaculate office and spread papers out, rapidly going through figures and details with Jim until his head was spinning. Neil's secretary came in quietly with a tray of coffee – no biscuits; Neil was a very austere and disciplined man – and left just as unobtrusively. The figures continued until Jim began to realise that he was no longer understanding very much of the detail. After about three hours non-stop, Neil stood up. 'I'm sure that's enough for you to be going on with, Jim. We'll let you get on. Let me know if you need any more information. I suggest you get your team together and meet up with John's team from Abbey this afternoon for an introductory session. I'll e-mail you a copy of these documents Jim for you to pull into shape for your team. Keep me in the picture.'

Jim found himself in his boss's outer office, looking blankly at Judith as she held a phone between neck and shoulder while she typed rapidly on her PC. She smiled vaguely as he managed to regain his composure and walk with as much dignity as he could muster across the carpeted room into the tiled corridor. For some reason he needed to at least pretend that he wasn't feeling totally shell-shocked and to get on with the job he'd been given.

When he got back to his own office he could feel several pairs of eyes boring through the windows that faced onto the main office, as he sat at his desk and concentrated on the papers in front of him. He worked his way through them, picking up a few more details, then started to phone his fellow team members.

'Hi Bill. Jim. Are you free for a meeting about the merger this afternoon?'

'Yeah, sure. What time?' Bill was a man of few words and always appreciated the direct approach.

'Let's say 2.30, my office. OK.

Next up was Ron. No reply so he phoned the switchboard and had him paged. A few minutes later, Ron rang.

'Hi Jim. How's married life?'

'Fine, Ron. Listen, you're on the merger committee with me aren't you? We're meeting in my office at 2.30. Can you make it?'

'Don't really know yet. Depends how it goes. We've got a lot on in the warehouse you know.'

Jim interrupted and headed off what he knew would be a familiar diatribe about Production always being behind and the disastrous effect that had on Ron's plans. 'It's important Ron.'

'OK, OK. See you at 2.30. Bye for now.'

He rang Caroline and agreed the time. She was a little inclined to gossip but she was a pleasant, efficient woman who knew what she was doing.

Knowing that Malcolm arranged everything via e-mail, he sent a quick e-mail and knew Malcolm would phone him if there were any problems.

He got his message through to Gerry Donahue, the Sales Manager at Martin's and Katrina Richmond, the Marketing Manager, via Gerry's assistant. They were both just leaving a meeting with a major customer but did not argue when they were asked to attend the Merger Committee meeting.

He tackled David Hardcastle, the Production Manager last. At fifty-eight, David had been coasting towards retirement for a while now and his defensive attitude caused problems for a lot of the other managers.

Hi, Jim. Have you heard about the merger yet?'

'Yes, that's why I'm calling. The first meeting of the Committee is this afternoon at 2.30. OK with you?'

'Yep, that'll be fine but I could do with a chat before that.'

'Sorry David, I need some time to get up to speed first.'

'Well, I wanted to get you on side about how it will affect Production. It'll be bad for Purchasing too, you know. Heads will roll.'

Jim knew that David would be worried but didn't have time to deal with that now. 'Don't worry, David, it's early days. Let's just get the ball rolling this afternoon.

At 2.30 Jim, David, Malcolm, Gerry, Katrina, Bill, Caroline and John Forsythe along with the whole of the Abbey Team were sitting waiting for Ron. The phone rang and Alison, Ron's Entry Clerk, informed Jim that Ron had gone up to the company's leased warehousing site about two miles away on an urgent job so he wouldn't be able to make the meeting.

Jim wriggled in pure embarrassment as John took control of the soon to be abandoned meeting. 'Well, I'll make this brief since Ron won't be joining us. We'll meet at 10 am tomorrow. We've got to look at every aspect of the business to see how we can make the merger work. Bring all the relevant details with you tomorrow.' He nodded to the members of his own team and then looked to Jim to take over. Jim grasped the chance.

'Caroline bring Department Head Counts. David, make sure you've got your throughput figures. Gerry and Katrina, could you use the time between now and tomorrow morning to get as much information as you can about customers, products, sales forecasts – you know the score. Bill and Malcolm, you know what you'll need, don't you?' They nodded. 'Right then see you in the morning.'

As soon as they left, Jim fired off a terse e-mail to Ron
From: HESWALL, Jim
To: WHITEHOUSE, Ron
Subject: Meeting
Meeting rearranged for 10 a.m. Tuesday. Be there. No excuses. Bring departmental information.
Regards
Jim

Jim spent the rest of the day checking the papers John Forsythe had given him, reading up on Abbey Products on the Internet and reading the dozens of e-mails that had arrived during his absence.

By 5.30 he'd had enough and headed out of the office. As he passed the vending machine, he dropped a coin in as he suddenly remembered that he hadn't eaten all day. He selected some chocolate and waited for his change to drop out. Nothing happened. A giggle exploded from him, then he looked self-consciously up and down the corridor to see if anyone had heard him. Who was it who'd said 'Change is inevitable except from a vending machine'? How very true.

As he drove home through the early evening traffic with the sun in his eyes, he pondered on the merger and the problems ahead. It suddenly struck him that he had a merger situation at home as well as at work. How would he cope?

Points to reflect on

How do you recognise change?

- Vital to recognise the fundamental need for change
- Find and define the imperative for change. Look for indicators that show the need to make a change. The cause in this case was a merger. Other indicators may include:

✓ Inadequate financial controls
✓ Poor customer service
✓ Losing market share
✓ Low growth – or no growth
✓ Excessive inventory
✓ Sales and cost issues

- Recognise resistance from any member of staff (in this case, Ron)
- Be ready to field questions on redundancy – but don't hide the possibility and probability

Chapter 2 – How to start

'I lost my job today.' Jim turned to look at Jan and immediately realised that he'd made a mess of telling her his news.
She put her arms around him and he could no longer see her face. She was talking and talking and he could hear the pain in her voice even while she was trying to reassure him that everything would be all right, that they'd manage. He felt the wetness on his face and pushed her from him.
'I'm not the Purchasing Manager anymore, I'm the Special Projects Manager instead.' He couldn't keep the smile from his face or from his voice.
Jan turned and ran. Seconds later he heard their bedroom door slam and he looked around him. He noticed the house was a mess as he raced through the living room and up the stairs.
'That was wicked.' said Jan as he sat down on the bed where she was sprawled, face down.
'Sorry, sorry, sorry. I'm not thinking straight.
'Mum,' yelled a voice from downstairs 'what's for dinner?'
Jan hurriedly sat upright and pushed her fingers through her hair.
'We'll talk later' she said as she stalked from the room. That sounded like a threat, thought Jim, I really should have thought that through. Why is life so complicated?
The evening meal was brief, disorganised and filled with chatter and moans from the four children as they sat around the dining table with their elbows touching and, at times, jostling each other. Jan ate her meal at the kitchen counter, washing up as she went along. As Jim went into the kitchen with the dirty crockery, Jan left. He finished the washing up and joined her in front of the television. Her stony face told him that she was still upset and the bumps, beeps and shouts

coming from upstairs told him that the boys were playing at some over-violent computer game.

'Where are the girls?' he said, pretending, as he usually did in these circumstances, that there was nothing wrong.

'Out.'

A few more minutes of silence as Jan stared at a soap opera on the television were as much as he could take.

'I'm sorry.'

'Have you any idea how that made me feel?' she asked, her voice rising and her eyes starting to look damp again.

'Yes, I...' he began

'No, you haven't. This is important. Do you know the failure rate for second marriages? We've got to make it work and we won't be able to if you don't think before you open your mouth. Anyway, what sort of a job is 'Special Projects Manager? Isn't that just a short stop before they show you the door?'

He started to answer 'Well, I...' but she cut him off again. When Jan got started, she took some stopping.

'A fella I worked with once was given the title of Projects Manager and he was made redundant within six months.' she ranted 'What are you smiling at?'

I'm smiling outside but dying inside thought Jim and there's no point arguing with her. Shouts from upstairs interrupted them and it was bedtime before they got back to the discussion. Then it was conducted in short whispers.

'So, go on then. Give me all the info.'

'Well, so far there's not much to tell. MML is merging with Abbey Products and I'm co-ordinating the whole thing for MML. I guess I'll be spending a lot of time over at Abbey too – you know, liasing with their team.

'And what will you do when that's finished? Will you go back to your proper job?

'Oh, Jan, I don't know and anyway I happen to think that this is a proper job. Neil Martin made it seem like a bit of a step up for me but now I'm not so sure.'

'Well, one thing's for sure – we can't afford for you to lose your job. Do you have any choice about this?'
'I think I want to do it.' he said, not sounding confident. 'If I make a go of it who knows where it might lead – it'll be a bigger company for a start.'
Jan changed the subject. She didn't think there was any point saying any more about the merger. Everyone knew that mergers didn't work. She didn't think that Jim knew what he was getting into. She changed the subject.
'We'll have to do something about the routine in the mornings. I was late for work today. I think we need to change a few things about the house. Wouldn't it be great if we could have another bathroom – just for the kids? Things are going to get worse, you know. The girls are both growing up. I'd like a bigger house really. It was fine when just the girls and I lived here. We had a routine but that's all been turned upside down now. I didn't realise how much extra mess three men about the place would cause.'
'Now hang on a minute Jan. Emma was the one who took half the morning in the bathroom!' exploded Jim. 'and anyway, I thought you didn't want to move house.'
'Keep your voice down. Well, what do you suggest?' came the whisper in the dark.
'It seems to me like I've got a merger at home as well as at work, so we need to think about what we're aiming at. Planning is the key and we have to involve everybody. We'll get the kids together tomorrow night.'
'Great, we can order Pizza for dinner.'
As Jim pulled into the car park the following morning after another fraught bathroom episode, he saw Ron Whitehouse talking to Caroline from HR just outside the warehouse doors. He went inside the main office building and Caroline caught up with him near the stairs.

'We've got problems there,' she said with a sigh ' he's convinced it won't work.'

'Will he be at the meeting this morning though?'

'Oh, yes. He can't wait to be proved right.' she chuckled.

The Merger Team members from Abbey arrived in several cars and as soon as everyone – including Ron – was settled around the large table, John Forsythe turned to Jim. 'Let's start the meeting. You go first with MML's details.'

Jim cleared his throat and looked nervously around the table. He'd never seen this room so crowded. 'Right, you all know about the merger. The brief for us is to go through all our departments – see what needs changing. Have you all got your information with you?'

Heads nodded around the table but still several of Jim's colleagues – and one or two from Abbey looked unprepared for what was to come, thought Jim. They have no idea about the size of this. Let's see what they've got.

'Caroline, what have you got?

'Well, here are the head counts.' She passed two sheets of paper over to Jim. 'To sum up, we've got an ageing workforce, no union presence to speak of, lots of experienced workers, HR have just had a new IT system installed. In fact, we're still struggling with it. Everyone's holiday entitlement was understated on last month's payslips.' There were mumbles and rumbles of discontent around the room. 'It's a very complicated database,' said Caroline defensively 'and, of course, we still have to run all the manual ways of doing things alongside it. There's one more thing I think I should mention. We do have a number of outstanding injury liability claims outstanding. There was another incident just last week. Karen from Quality Control fell in the Finishing Room and injured her back. I think she fell over some stock that should have been scrapped or something. It's a problem that's increasing.'

'David, any comments?'

Yes, there is a problem with accidents but I don't think it has any connection with anything we're discussing here.'

Bill stepped into the discussion for the first time and was unusually fervent. 'On the contrary, finding the right ways of working can solve all sorts of problems and that seems to me to be why we're all here.' He looked quickly at Jim for agreement. Jim nodded in encouragement and Bill continued. 'If you knew where all the stock was – and it was in the right place, of course, then some of those accidents definitely wouldn't have happened.'

'That's something we'll have to look into.' said Jim scribbling on his pad 'Anyway, David can we have your view on the current status in Production?'

'I don't think it's any secret that I've been unhappy with the way things are going for some time.'

Jim looked at the Production Manager and nodded. 'In what way?'

'I just don't trust the info. And we're all working with different software packages. We have to have meetings to find out what's going on. Surely it's possible for us to have an IT system that gives us reliable details of orders outstanding, work in progress and so on?'

'Well, yes it is. If course it is.' interjected Malcolm. 'Not with our current software though.'

David resumed. 'The other thing is that Work in Progress is continually increasing. We just can't cope. Neil has been complaining about us having so much WIP. I thought employing the new progress chaser would help but the totals have carried on going up. Of course, the Sales Department are always changing their minds.' He glared at Gerry who opened his mouth to retort but was silenced as Jim leapt back into the conversation, anxious to avoid a confrontation at this early stage. He could see already that there were plenty of issues, just bubbling under the surface.

'OK, David, I see. Ron, you next. What systems are you running in the warehouse?'

Ron cleared his throat and Jim's heart sank. It was going to be the usual. 'We work in a unique way. All the stock records are manual. We know exactly what we've got of every item of stock. It's always late coming to us though so it's a miracle that we get to enter it into stock at all before it's needed for...'

'For heaven's sake Ron, give it a rest. ' interrupted David Hardcastle, the Production Manager '

'Don't you have any computerised stock control systems then?' asked Jim.

'Oh, yeah,' said Ron 'but one of the girls keeps those up to date. The records at the stock locations are what matter. I've brought you the computer printouts but you should check the actual records too, you know.'

'Do the manual and computer records agree?' asked Jim

David grinned but quickly covered his face with his hands.

'No, they can't do, can they?' said Ron with a straight face as though he'd just said something sensible.

Deciding that now wasn't the right time to try to change Ron's habits of a lifetime, Jim moved on to Bill who handed him a two inch thick pile of computer print outs.

'You should find everything you need there on the finance systems but there are a lot of other records and reports held manually. Let me know if you need some of the detail about the P&L figures or to go through anything in more detail. The stock figures – all of them – the finished stock in the warehouse and Work in Progress and Raw materials too have risen sharply over the last year or so as you will see. Also, we're selling more in terms of volume but the profit margins are decreasing. Those underlying problems need to be addressed quickly. I've got one or two ideas about software systems and so on – maybe we could get together.'

'Well, Malcolm needs to be involved there. You two OK for this afternoon?'

'Sure,' said Malcolm ' here's my contribution re the software in use here. I'd bet that Abbey's will be totally different.'

'It's to be hoped they have some useful information on progress chasing – ours is worse than useless.' said David as he shot a look at Malcolm.

'Good IT systems cost money and you know we just don't have the budget.' Malcolm looked under pressure as he attempted to field the criticism so Jim moved swiftly on.

'Let's have a look at Sales next. Gerry?'

The tall, tanned young man stood up and walked briskly round the table, handing out a brief summary of the sales department at MML. 'These should help people to see how well we're doing and give you all the information you need. Jim looked down at the slickly produced presentation. The thought that it looked good but didn't have much substance chased through his mind but he thanked the Sales Manager and indicated that Katrina should sum up the marketing situation. After that he passed the meeting over to John with a barely-suppressed sigh of relief and listened patiently as the Abbey people went through the details of their departments. As lunchtime approached, John called the meeting to a close and Jim turned to the MML team.

'Right then, I'll get on with putting this together and I'll see you – Bill – and Malcolm at 2.30 back here. Thanks everybody.'

That hadn't gone as well as he hoped. There were obviously a lot of issues under the surface and the fear of job losses was apparent in everyone's defensive attitudes.

Left alone, Jim made notes on what he'd learnt from the meeting and added his own notes about the purchasing function. In his purchasing role he had ordered what people told him was needed but often realised that nobody really knew what was needed. He hoped that whatever came out of the changes, a better forecasting system for raw material requirements would be on the list. He quickly came to the conclusion that MML had problems. The processes being used to run the business had to be questioned and the figures they produced just weren't reliable.

In the afternoon, Malcolm and Bill filed into Jim's office, Bill carrying a huge pile of brochures and files. Jim dreaded this. How much more of this detail could he handle?

'Right Bill, you mentioned new software, what have you got there?'

'I've been looking at ERP systems for a while now and...'

'You've no chance.' interrupted Malcolm

'Hang on.' said Jim 'Go back a bit. What's ERP?' Little did he know that before the merger process was over, he would get to know more than he ever dreamed possible about ERP – and probably more than he wanted to know.

Bill took a deep breath. 'It stands for Enterprise Resource Planning. It's a method for planning and controlling all the resources a company uses. There is software that enables it to work on a single computer system. It can also be web enabled. All the bits work together – a bit like the way the different parts of Microsoft Office work together – Excel, Word, PowerPoint, Publisher and so on. That revolutionised how people work. ERP could be our revolution. We certainly need it to pull everything together but I've not managed to convince Malcolm.'

'I'm not arguing with you but we just don't have the budget. And just remember you're dealing with people like Ron who thinks Resource Planning can be done with just a holiday calendar!'

'Yes, and people like Neil who thinks technology is nothing to do with his problems here. I understand your problems, Malcolm and I'm not having a dig at you but...'

'So, I assume you'd agree that MML's systems leave a lot to be desired.' said Jim. He knew from his own experience that the company's systems didn't really help him in things such as reviewing suppliers and he still wasn't happy with the purchasing policy even though he'd started to implement it over two years ago.

Bill nodded his head. 'That's right. As I see it we have three main problem areas – poor customer service, competition causing pressure on margins and inadequate management information. The consequences for the business of those three problems are horrendous. We have late deliveries all the time and customers complain and then they cancel their orders. This leaves us with stock that we can't sell to anyone else or we have to re-book the delivery. If a customer turns away a late delivery – and these days many of them give us just a fifteen minute timed delivery slot – it means we have to bring the goods back, double handle them, store them and then, if we're lucky, the customer takes delivery later. I don't think our stock recording systems can cope with those problems. It's certainly difficult to reconcile the stock figures every month.'

Malcolm spoke up. 'And I'm always in trouble with the month-end printouts. They're always late because of all the queries on the system and constant changes demanded by department heads.'

Bill's enthusiasm for the possible changes was clearly visible. 'Yes, most of the time they come out so late – by the time I've added in the bits that I have to do manually – we're well into the following month and so nobody really takes any notice of the figures. Plus, we have a high percentage of invoices unpaid but we don't seem to be able to sort out why. The most important thing though is that poor Management Information Systems lead to poor business decisions. We must take this opportunity to put that right.'

'This ERP stuff sounds good to me. Let me go through these.' Jim pointed to the brochures.

When Bill and Malcolm had left Jim got down to business. First he read through the details that Bill had left with him then he made himself a list. I'm learning from Jan on this one, he thought, she's always making lists. On one side of his sheet of paper were the problems that they needed to solve. He hoped that the solutions would come to him later. As he went through the many problems that had been identified to him since he began as Project Manager, the reasons for them also occurred to him so he jotted those down too. When he had finished for the day he had plenty to think about. His list was as follows:

- Customer Service Department – too many queries (customer complaints!)
- Suppliers not being paid (not automated, errors in input)
- Late deliveries from suppliers (see above!)
- Production complaints re raw materials
- Over ordering of raw materials (assume precautionary measure by David)
- Too much stock of finished goods – complaints from warehousing
- High outside storage costs (too much stock/not enough space)
- Long lead times for non-standard items
- Complaints from Sales re lead times
- Poor Sales Forecasting (where do they get their information?)
- Sales Department complaint re stock of old, unwanted items

- Customer complaints re old, dirty, damaged stock sent out
- Falling sales figures (customers going to competitors for new products)
- Falling Sales bonuses
- Purchasing Dept working with poor forecasts
- Untidy factory
- Poor information

He put down his pen and stared at the list. It was obvious! This lack of integrated information was affecting the whole company. A drastic solution was needed – for the good of the entire company. Every part of it was affected. He worked through his list again and saw that, from what he had already learned about ERP, that could be just what the company needed – possibly the only solution.

As he drove home, his thoughts turned back to the problems with the house. They weren't so different from his problems at work. Jim pondered that maybe, just maybe, the problems both at home and at work could be problems that would lead to opportunities. He had another 'merger meeting' tonight. He supposed this would be similar – everyone will want something different and the ideal solution would be out of his grasp. Still, nothing ventured, nothing gained, he thought and at least we'll get Pizza at this meeting.

Points to reflect on

Identify key requirements from the business

- Look at what is the benefit of the change
- Assess the value to the company of the change
- Define what ERP is – and what it is not

* A useful definition of ERP is supplied by APICS:

'A method for the effective planning and control of all the resources needed to take, make, ship and account for customer orders in a manufacturing, distribution or service company. It is not a piece of software – software is an enabler.
ERP provides an organised communication system so that high level operating philosophies and strategies are followed during the tactical operations of the business.'

- Other indicators for change include bad information, lots of queries for information both internally and externally
- Recognise that all functions are inter-related
- Envision the business as a system of processes not individual departments

* Reproduced by kind permission of APICS
APICS Dictionary, 10th Edition, © 2002

Chapter 3 – Define the boundaries

'Right, what sort of Pizza does everyone want?' As soon as he said this, Jim regretted it. The answers came thick and fast.
'Ham and Pineapple'
'Something spicy'
'Just a plain one, please'
'Do we have to have Pizza?'
Hey, hold on,' said Jan 'just how much is this going to cost?'
'Don't worry, I'll get an ERP type'
'Doesn't sound appetising. What's that?
'Wait and see. You'll like it.' said Jim with a smile.
Forty minutes later, the giant pizzas were on the table. As she added glasses and paper napkins, Jan had a peek at the Pizzas. Ah, that explained it. Jim had ordered the ones with four different flavours in one pizza. What did ERP stand for though?
'ERP stands for Enterprise Resource Planning and it's a system to serve all purposes – even when we all want a different topping on our pizza!'
Five blank faces met this announcement. The four smallest ones gave up trying to figure out what Jim was talking about and tucked into their pizza. Jan, meanwhile looked at her husband with a grin.
'I suppose this will be more of a merger meeting than a family discussion, then?'
'Just eat your pizza, woman. I'm the Managing Director around here.'
'No, you can't be,' said Jan with her mouth full 'you're in a very secure position as Special Projects Manager!'
As soon as the table was cleared, Jan called the meeting to order.
'Before we start to talk about the house, I just want to ask, who used the last of the soap?'
Nobody answered.
'Could anyone who uses the last of anything – soap, cornflakes, whatever – please write on my new 'we

need' list on the front of the fridge door. That way I don't have to write another shopping list and you will be able to see whether someone else has already thought of it. I have already put regular items on the list so you don't need to remember or add those. It seems that I've been appointed as Inventory Manager for this newly merged organisation and I'm determined to make the job go as easily as possible.'
She paused and looked around at the bemused faces. The boys nodded, their expressions serious. 'Now then we've decided that the house isn't quite right for us so we want to ask you all what you think it will take to make it right. We need to work out how we can make the most of what we've got or perhaps think about moving to a new house, how we can avoid the rush for the bathroom in the mornings and how we can all live together without fights and arguments.' she announced. 'And, very important, how we can make sure we all get to work and school on time? If we can't sort that out, then we won't be thinking about a bigger house, that's for sure. We'll be looking for new jobs!'
They all started to talk at once and Jim intervened, raising his voice to be heard above the noise.
'Let's start with James.' All eyes turned to the eight-year-old who was sitting on Jim's left.
'I think we should have another computer. Andrew is always using ours. It's not fair.'
'That's not quite what we...' started Jim but Jan silenced him with a look and wrote 'Computer' next to James' name on her pad.
'Right, Andrew, what do you think about the house?'
'I agree with James.'
'That makes a change.' muttered Emma
'Yes, I do. One computer between two people isn't enough.'
Jim leapt up and got himself some paper and a pencil from the sideboard drawer behind him. He started making his own notes on the meeting.

'Well, what else do you think we need – all of us?' asked Jan

'A swimming pool.' said Andrew triumphantly. 'Write it down, Dad'

'OK, but we need to think how life is now and how we would like it to be. Emma, you're next. What do you think we need to change?'

'We need another bathroom. One isn't enough for us all in the mornings. Someone's always shouting at me to come out when I'm not finished.'

'They wouldn't have to if you didn't try to make yourself look cool for school every day.' shouted Andrew.

'That's enough, Andrew.' said Jim

'Well it's true, you said so yesterday!'

Refusing to argue with his ten-year old and attempting to cover up at the same time, he moved rapidly on.

'Philippa?'

'I'd like my own bedroom please.'

'Jan?'

'Well, I think we need more space all round. I'm going to make a list of all the things we do in the house and what sort of problems we have – such as the bathroom situation. I'd like an ensuite bathroom – or perhaps a shower room -for our bedroom and a bit more space downstairs.'

'Perhaps a conservatory would give us some space.' suggested Jim 'and the bathroom is a definite problem area so that goes on the 'wish-list'. Everyone wants something different so we need to decide what's possible and how we can go about getting it. One thing's for sure, I won't have time to organise it along with everything that's going on at work. You're too busy too, Jan, so if we do decide to do any extensive work, perhaps we should get an architect in to oversee the project.'

'Can I go and play out now, Dad?' said James, sliding off his chair and racing for the freedom of the back garden.

'Will we get a swimming pool, then?' asked Andrew, hope shining in his eyes.

'I don't think so, Andy, not just yet, but we'll see.'

A few days later, Jan and Jim sat down again at the dining table. Philippa was upstairs listening to music and the other three were out doing various activities – swimming, music lessons and football practice.

'You're a great one for your lists, aren't you? Are you making a bid for the Special Projects Manager's job?'

'Well,' said Jan 'it pays to be organised. I've made a note of what the current problems are, what we use the house for, our wish list – including Andrew's swimming pool – and what I think we should realistically aim at, perhaps get some quotes for an extension or see what's up for sale locally and so on. I've also made a note of a few architects and builders but I don't know anything about any of them. Do you know any architects?'

'No,' he said 'but I'll ask around. You never know. The boys don't seem interested, do they, especially if they don't get what they want. Balancing wants and needs is going to be difficult.'

'Paying for it might be even more difficult.' said Jan 'We do need to make sure that all of them participate though. Any ideas?'

Just then the doorbell rang and Jim jumped up to see who it was. As he hurried through the hall, he stubbed his toe on a Skateboard that one of the children had left propped against the wall.

He gritted his teeth and advised the smart young man at the door that no, he didn't want to buy any brushes or cleaning products and limped painfully back into the dining room.

'Life would be a lot easier around here if everyone put everything away.'

'You're a fine one to talk. You don't seem to have discovered where the laundry basket is yet.'

'Sorry. I am trying to be tidier – and I'll have a word with the boys.'

Jim used his time on Monday in the office to offload his Purchasing duties to his assistant and to move his files and notes on the merger and MML's systems to a small office between his boss Neil's office and the boardroom. He also made a start on notes on MML along the lines of Jan's notes at home. Jan's right, he thought, it does focus the mind, this list making.

On Tuesday, he was first in the bathroom as he made an early start for the meeting at Abbey Products. He didn't really know what to expect as he walked into the rather plush reception area flanked by flags of various nations, with impressive-looking certificates and mission statements on the immaculate, cream-coloured walls. The place even smelled fresh and clean. A uniformed receptionist greeted him and, as he introduced himself, it was obvious that she was expecting him.

'Yes, Mr Heswall, Mr Forsythe's secretary will be down for you shortly. Would you like to take a seat?'

Jim took the opportunity to have a look around the reception area, to get a feel for the company. Even at 8.30, the company was buzzing. Calls coming in, staff arriving or walking through the reception area and another visitor already waiting. A smart, middle-aged woman in a navy blue suit approached him.

'Mr Heswall? Could you come with me, please. The meeting will be in the board room.'

He followed her along the carpeted corridor to a large room where he was introduced to the only other occupant.

'Mr Heswall, meet Neville Crosland our IT and Logistics Director.'

'Jim Heswall, pleased to meet you.'

John Forsythe and Tony Abbey entered the room as the secretary returned with a tray of coffee and Tony launched straight into the meeting with what Jim was to realise was his customary enthusiasm and energy for the change project.

'There's only one item on our agenda today – getting our two companies to work as one.' Tony Abbey looked directly at Jim. 'All my staff know that I'm one hundred percent behind this merger, Jim, but there are some real problems to be solved before we can make the most of the opportunities that it presents. We won't do that if we keep working as two separate companies.'

Jim decided that he must start as he meant to go on and plunged in with a question. 'I'm sorry to interrupt, Mr Abbey but...'

'Call me Tony, please.' said the MD with a smile.

'Er, right, Tony, Can you clarify what you mean by working as one company? Do you mean all our people working together in one building?'

'That's just one of the things we need to find out. The changes will have to be made in lots of areas. We need to find synergies, get the companies working together, make culture changes, ensure that we have effective communication throughout the company, sort out the direction. Each company has lots of different ways of working and there doesn't appear to be much communication yet. We absolutely must make real changes. The Merger is just the driver of those changes and we need to set out our strategy for the future. We want to be the best in our industry.' He looked around at the attentive faces and, getting really into his stride now, rushed on. 'We must make sure that our customer service is the best it can be, that we get our marketing absolutely right, and do what our customers want us to do. We need to find the best way of working. Getting – and keeping - the right staff is so important. Our people must want to work for us. If we create a motivational culture we will attract the best people. And our technology - that is key. It must be leading edge. Tested and proven, of course, but the absolute best available.' He beamed around the room and Jim could almost feel the energy and depth of feeling that Tony displayed. 'John here will tell us what he has planned to kick start the change programme. But first, perhaps you could sum up the IT systems in place at MML?'

'Well, we've got a variety of systems, most of them are PC based and we have a mainframe where our financial records are dealt with – invoicing, credit control, management information, that kind of thing.'

'John, that's what you expected, right?' said Tony

'Yes, and our systems are a similar mixture. The usual PC based databases, off-the-shelf software as well as the specific stock bar-coding system in Production and Logistics. That might be a system we want to retain, I think.'

'The question is, do we have the competencies in-house to get all our systems in line?' said Tony

John handed round a single sheet of paper as he said 'I've prepared this competency matrix for Abbey but as you can see, it's not promising. As you know, I've plenty of experience in this field and my opinion is that we really should be looking at installing an ERP system – perhaps with consultants to help us.'

Jim brightened, sat up straight and said 'It seems we're thinking along the same lines. I've got a few details about ERP software here.' He brought out the notes and brochures from his meeting with Bill.

'That's great.' John Forsythe leaned over and grabbed one of the brochures saying 'I thought we might have a bit of an uphill struggle converting you MML guys to ERP but you seem to be ahead of us.'

'Do you by any chance have any experience of using any consultancy firms? Any preferences?'

'No, I'm afraid I don't.' said Jim, relieved that he was managing not to appear totally out of his depth and resolving to buy Bill a pint and have a chat with him as soon as he could.

'Neville, what's your opinion?' asked Tony

Neville cleared his throat. 'It's the best way of pulling our two companies together. We can get some real benefits from the merger if we invest in consistent processes.'

'I don't have any direct experience of consultants' involvement in this type of project but it certainly won't be cheap.' Neville continued.

Tony nodded. 'We need to find the money to do this properly. If we duck out now, we might leave ourselves with no choices whatsoever. What other high-cost projects have we got outstanding, John?'

John did not need to consult any notes. 'We've a number of expenditure requests currently under consideration. There's the plans to renew the transport fleet at MML, the Abbey launch of our product range in Japan – both of those have been put on hold, of course – and there's also the request from Marketing for funding of the PR campaign to promote the new company.'

'OK' said Tony 'Nothing there that can't wait until we've decided the expenditure on consultants, new software, the whole change programme. Right, before we get any consultants involved, John, we will all need to be clear about our strategic direction. You and I need to get together with Neil Martin – and you, of course, Jim. I'll get my secretary to set up a meeting over at MML.' With that, Tony stood up and looked around the table at the three men. 'Sounds like I can leave it with you then for now. I guess you and Jim are going to have to work pretty closely on all this, John. Keep me in the picture when you've got a budget together for this plus a preliminary plan of action.' He shook hands with Jim. 'Give my regards to Neil when you get back won't you? I'll see you soon at the Strategy meeting hopefully. Goodbye'

With that he was gone and John turned to Neville and Jim. 'Tony's right. We could get to know one another and carry on with this over lunch. How are you fixed Neville?'

'Sorry, no can do. Meeting at 1.30.'

Jim breathed a sigh of relief as Neville left the room. John seemed not to notice, picked up a sheaf of papers and said 'Might as well walk down the road to the local pub. Reasonable food, fairly quiet, let's go.'

'Well, what do you think of Tony?' asked John as soon as they were clear of the building.

'Decisive, clever, has a reputation for being a shrewd operator. I liked him. Do you think I'll feel the same way when all this is complete?'

Probably,' said John ' he's a good friend as well as my boss. He headhunted me from my last job and he really is on board with this merger. He'll do what it takes to bring the two companies together. You're not quite so keen on Neville, are you?'

So much for John not noticing, thought Jim, he had a hunch now that there was very little that got past John. 'Well, no, but then again I hardly know him – just a first impression that he's not too keen on the merger.'

'You're probably right. I think he's frightened of all the work he knows will be coming his way.'

Jim looked closely at John as he asked 'Tony said there are problems. Are they so serious?'

John looked back at Jim just as closely 'They're worse. As Tony said, changes are inevitable. The fact that we've at least five legacy systems that don't communicate means that we can't allocate costs, we don't have an accurate figure for stock at MML and we have no chance of even maintaining sales figures in the future. We certainly can't expect to increase them until we get our systems in order. And that's just the tip of the iceberg. Tony knows there is some hard work ahead of us but you can bet if he's put his money into the venture, then it's possible to sort it out and make better profits in the end. You were absolutely right about his reputation as a shrewd operator.'

They walked into the bar, ordered drinks and picked up menus as they seated themselves in a quiet corner.

Jim took a long drink and said 'That's better, it's thirsty work pretending you know what you're doing.'
'Don't be so modest. You seem to be up to speed. I've had a bit of experience in this sort of thing.'
He didn't elaborate and the conversation turned to more personal matters after they had ordered their meals. 'Funnily enough, you mentioned experience in mergers,' said Jim as he tucked into his steak with enthusiasm, 'I'm going through another merger situation at home just now and it's amazing how many similarities there are. I remarried a few weeks ago and with two children each, we're finding it a bit of a squeeze at home. Mergers are difficult, aren't they?'
'I'll say. Are you going to have to move house?'
'Not decided yet. We're looking into extending at the moment – an extra bathroom, extra bedroom, conservatory perhaps.'
'Oh, you have my sympathies then. We had to build on to our house when we had triplets – but that was over twenty years ago. A similar situation to yours, I suppose.'
Jim laughed and said 'but that was more like organic growth than a merger!'
John joined in. 'Yes, it certainly was but it caused just as many sleepless nights! Look, we're going to have to work together on this company merger, why don't you and your wife come to dinner sometime soon and you can see the work we had to do on the house. We'll soon be starting again – with the building work, not the triplets I mean – because Brenda's mother will be coming to live with us I think, so we'll be building a 'Granny Annex' next. Couldn't stand to have the old girl living in the same house as us now that the children have left home and we've finally got a bit of peace and quiet– but for heaven's sake don't tell Brenda I said that. There's a skill to managing building work, just the same as with managing business change, you know.'

'Yes, that had already occurred to me. I'm not sure that I have the know-how for either.'

'It's all in the planning. There are a few definite stages to the process – whether it's families and home alterations or companies and business change.'

'I've already taken my wife's advice on planning at home – she's always making lists. I've started noting where we are now, our wish list and realistic expectations.'

John grinned. 'Ha! Just like business change – I'm sure that in the weeks to come we'll be making similar lists but we'll be heading them 'as is', 'could be' and 'to be'.'

Jim was enjoying his lunch with John but reluctantly looked at his watch. 'We'd better be getting back, I think. No time for dessert.'

'Just as well.' said John patting his trim waistline.

They strolled back to the Abbey building in the spring sunshine and resumed their business discussions in the boardroom.

John returned to the subject of choosing consultants. 'I've heard of a couple of local consultancy firms with good reputations. I think we should ask them in for initial discussions along with one of the big boys in the consultancy world. I've worked with one of the biggest – WPC – but I wasn't impressed.'

'Will you get in touch and let me know when you want me to come back for the meetings?' asked Jim

'Sure.'

John accompanied Jim down to the car park. 'I'll ring you soon to sort out a date for that dinner we mentioned. In the meantime, I'd say that you should learn everything you can about ERP – it'll come in very handy.'

'I certainly will.' said Jim turning towards his car ' Oh, by the way, any advice or recommendations about finding an architect?'

'Mm, just the same as business consultants – get someone with a good reputation who you feel you can get along with. Get quotes from two or three, then go with your instincts. As simple – or as complicated – as that.'

Jim watched him walk briskly back in to the building and then started his drive home. He was well satisfied with his day's work and started planning how to get everyone involved – both at home and at work.

Points to reflect on

- Priority should be on business change and how this will work – not just looking at a new system or software
- It is useful to identify competencies at an early stage.
- Look at what other projects may be going on. Will ERP be the priority?
- Recognise the need to offload some/all current responsibilities if working full time on the project
- Decide what is included in the change/ERP project and, equally important, what is excluded. Maintain constant reference to this or the project will grow!
- Get executive buy-in from the start

Chapter 4 – How will the project work?

'What's that?' asked Jim as Jan got out a large, bright blue, hard backed notebook. They were back at the dining table again with all the children assembled after their evening meal.
'I've decided to keep everything about the house project in one place – like a diary, I suppose. I'm going to keep all the phone numbers here and all my lists, what's agreed, comments etc. It'll be handy.'
'Right, sounds good. Did you have time to ring any architects or builders today?'
Yes, there are two coming round on Saturday, so make sure you're here. By then we need to have our ideas sorted out about what we want them to quote on.'
'Great, I'll be here.'
'OK, we're looking at getting quotes and ideas for a bathroom, a loft extension to give us an extra bedroom, and a conservatory.'
'What no swimming pool?' said Andrew
'Maybe not this time, Andrew' said Jim
'Well, do I have to stay here. Can't I go and play on my computer?'
'It's not your computer,' squealed James 'it belongs to both of us.'
'Quiet, guys, you have to stay for a few minutes. This affects all of us.'
The two boys leaned back on their chairs, thunderous looks on their faces. The girls were also starting to look bored so Jim quickly carried on.
'Girls, what do you think about the plans so far?'
'Great, who'll have the new bedroom?'
'That's not decided yet but we'll all be better off. We'll have more space downstairs if we have a conservatory and a bathroom will be wonderful won't it?' said Jan, trying to get the girls involved. 'All of those things should be possible whether we go for the new house or an extension.'

Jim realised he needed to back her up. 'There'll be room to play in a conservatory, even when it's raining and if it means I won't be late for work, the new bathroom might just save my job!'

'And mine,' said Jan 'and that'll be better for all of us.'

Jim could see that the boys were far from convinced. 'We just don't have the space in the garden for a proper size pool, you'll be able to keep going to your swimming training at the pool in town.

'OK, that's cool' said Andrew but he didn't look totally convinced.

The phone rang and Jim answered it. 'Oh, hi John. Great to hear from you. Yes, Saturday will be fine I think. Let me just check with the boss.' He looked across at Jan. 'Are we OK for dinner with John and Brenda Forsythe this Saturday?'

'I think so, so long as Melanie can look after the kids for us. I'll check with her. Can we ring him back?'

'Seems OK, John. Can we confirm later? Yes, sure, bye.'

'OK folks, we'll finish up here, you can all go and do your own thing.' said Jim as he got back to the table. No point keeping the children any longer. He hoped that he would be more successful with the staff at work but he doubted that it would be easy. Just like the family, there were people who wouldn't be 'on side' and, of course, everyone wanted something different. The trick would be in creating one team.

The next day, having first briefed Neil about the meeting at Abbey Products and advised him that hiring a consultant seemed like the way forward, Jim went to see Bill in Finance. Neil had not been impressed. He was a little old-fashioned in his approach to business, thought Jim, no wonder we've got a mish-mash of systems and not enough PCs to go round. It hadn't escaped Jim's attention that many people didn't have a PC on their desks. He didn't really know how they managed without one but no doubt he would have to find out. His thoughts turned back to home. Perhaps Andrew was right. He resolved to see if they could squeeze money – and space – into the budget to get the boys another computer.

'Bill, your idea went down well. It looks like we'll be getting consultants in to help with an ERP programme.'

Bill looked slightly embarrassed but pleased.

'Tell me what you know about ERP then. I need to get up to speed fairly quickly.' said Jim

'Well, I've a couple of books you could borrow. It's quite a long process you know, and there'll be a lot of work to do in terms of getting everyone involved and working in the same direction. It won't be cheap either – although there will be definite benefits – are you sure that Neil will go with this?'

'Not really, although I don't think he'll have much choice. Tony Abbey is really behind it and he seems like a determined character. Who else do you think will be a problem here?'

'Well, Ron Whitehouse in Logistics, obviously, but I think most people here will welcome some improvements in IT – and Malcolm won't be able to believe his luck if it all goes through. He really struggles with Neil for extra spending but he never gets anywhere.'

Jim's mobile phone rang. It was Jan.

'We're OK for Saturday with the Forsythes – Melanie will baby-sit. Can you let them know?'

Jim rang John at Abbey Products as soon as he left Bill's department.

'Just to confirm, John, Jan and I are OK for Saturday. Could you e-mail some directions for us? Thanks. Any progress with the Consultancy firms?'

Jim hung up after a brief chat with John who advised him that two meetings with local consultancy firms had been set up for the Monday of the following week.

On Saturday morning, Jan answered the doorbell and showed a rather large man into the house, introducing him to Jim as Barry Bryson, Architect. Jan and Jim showed him round the house and garden and after about twenty minutes, they all sat down at the dining room table. During the tour of the house, Barry had taken a few measurements but he'd spent more time chatting to Jan and Jim about what they wanted from the house.

'Well. I can see what the problems are,' said Barry 'and I've a few ideas for solving them. I'll put my quote in for a loft conversion because I think that's the way you will get the best use of the space available. We'll include a conservatory and perhaps a bathroom downstairs – I think we can squeeze that in under the stairs. I'll set out a timetable for you as well as the costs. Of course, if you want to look further into a new build project, just let me know. I could probably oversee that for you and I know some good people in the business.'

As soon as he'd gone – he was with the Heswalls for over two hours – another man appeared at the door. He was a small, quiet man dressed in workman's overalls that were extremely grubby. After their tour of the house, Jan made sure that the meeting took place in the kitchen. She didn't want this man sitting on her chairs. Luckily he'd taken off his work boots by the front door but his overalls looked just as dirty.

After he'd gone. Jan couldn't contain herself. 'Barry seems to know what he's doing. Doesn't he? And we certainly don't want that nasty little man in the dirty overalls doing the work. I can hardly wait for the work to start.'

Jim looked amused. 'I hope you can keep up that level of enthusiasm when the work does start. Don't forget that we've that architect coming for a chat tomorrow. You haven't given up on the idea of a new house have you?'

'No, I dream about that too.' she said

Jan's enthusiasm increased later that day when she was shown around the Forsythe's house. It was beautiful and Jan could feel herself getting carried away with plans for new carpets and curtains and for how she and the family would use the extra space. Jim brought her back down to earth by pointing out that the Forsythe family didn't include four school-age children.

'How did you manage all this, Brenda?' said Jan, determined to make the most of the opportunity she had to get the advice she needed to make this project work.

'I think we just muddled through but I would say that you definitely need to get an architect involved if the project is anything bigger than just decorating. He'll be able to make sure that everything happens in just the right order. We had a horrendous experience with a builder who tried to cut corners. He kept trying to lay floor tiles before the cement was dry so the whole thing had to be done again and he actually installed kitchen worktops before the plumbers had finished so that some of the cupboards had to be moved to get to the pipes. We had to get rid of him and get an architect involved at that stage because it was obvious that the work just wouldn't get done without someone to oversee the project, getting everyone involved and doing everything at just the right time'

Meanwhile, the men were discussing consultants.

'Yes, the two local firms have agreed to come in on Monday for an initial meeting and I think they'll be able to give us some ideas and a quote quite quickly but the other firm I approached – National Consulting – have said they'll get back to me in a couple of weeks with some suitable dates for their first visit.'

'Not in a rush then, are they?' said Jim, who really wanted to get on with this project.

'Well, let's see what comes up on Monday.' said John

By the end of the evening, both couples were sure that a friendship had started and Jim and Jan agreed to set a date for the Forsythes to visit them for a meal soon.

On the way home, Jan wanted to discuss the building work while Jim was a bit distracted thinking about the meetings with the consultancy firms coming up.

On Monday, Jim entered the Abbey Products reception area and was immediately told to go up to the boardroom. Again, he was impressed with the reception staff and pleased that they recognised him already.

John joined him walking along the corridor to the boardroom.

'We've one firm in at 10 a.m. and the other at 2.00 p.m. so we've an hour or so to prepare. Let's go through some of the details of MML's systems that you left with me last week.'

'Sure. What sort of thing do you think these consultant guys will want to know?'

'Oh, they'll need all sorts of details – they won't hesitate to ask for whatever they need – and it will be interesting getting all the details together. We've a lot to find out about each other's companies so this will be just the start of a very steep learning curve.'

'OK, let's get started.'

They worked solidly until John's secretary interrupted them to advise that the first visitor had arrived.

They were joined in the boardroom by Tony Abbey and Neil Martin who Jim was a little surprised to see there – he hadn't realised that the meeting was going to be on that level. He started to wonder what sort of a budget they were looking at for this whole project, How much would they have spent by the time the ERP software was up and running?

The visitors were shown into the boardroom – two middle-aged men and a woman who looked to be somewhere in her thirties – and John led the introductions.

'This is Tony Abbey, MD of Abbey Products - he's the Sponsor of the project – and this is Neil Martin, MD of Martin's Milling.' John nodded towards Jim. 'Jim Heswall is heading up the project on behalf of

Martin's. He's the vital link between the two companies.' Jim smiled around the table and felt that his position had been subtly talked up by John but wasn't about to argue.

They discussed general details about the merger first and then got down to the real issues such as IT, Sales, Marketing, HR and strategy. After almost two hours, the consultants seemed to have completed their initial sales pitch and fact finding and said they had sufficient detail to prepare a quote but would need copies of documents describing the merged company's strategic plans and goals for sales, marketing, technology and HR along with financial targets for costs and profits for the next few years.

Graham Manders, the leading consultant of the three who had been doing the presentation, summed up. 'As many of those details as you can possibly manage to supply please – it will really help us. The next stage will be a three-day Scoping Workshop. We'll lay out the details and requirements for that in our offer.' He emphasised yet again the number of high-profile clients they had worked with and also the need for extensive preparation before commissioning new software. Any questions?' The intelligent-looking consultant sat back and waited for their response.

Neil was quickly on the attack, asking about costs. Predictable, thought Jim.

'We can't estimate without a lot more detail of course, but it does seem like an extensive project and final costs will depend on the software selected, the amount of training necessary and so on. But, of course, the savings and increased efficiency will more than cover those costs. Otherwise, we wouldn't go ahead.'

Neil did not look pleased with the answer but said no more.

At this point the consultants left with a promise to be in touch before the end of the week and the rest adjourned for lunch. Jim took the opportunity to get a few more details from John.

'Sounds good, but what is a 'Scoping Workshop'?'
John grinned. 'It's 'consultant-ese' for a brainstorming session. A lot of information can come out of them. Gives us a chance to see the consultants' proposals in more detail and whether they really understand our problems. Also, it makes it clear who's on side and who's not. Mind you, that is becoming clearer all the time anyway. Tony, me, you – we're all well behind the project. Neil and Neville, well' he paused 'well, they've got issues of their own, I think.'
'Yes,' said Jim 'Neil is worried about the money obviously. That's why we haven't got up to date technology at MML already. He's always watched the budgets very carefully. I don't know what Neville's problem is though, do you?"
'I think he's feeling threatened.' said John 'It's very common in merger situations. Something we need to keep a look out for in all the different departments. People will be very worried about their jobs – especially when it becomes known that we're bringing consultants in.'
Jim looked worried. 'Yes, it's something that has certainly crossed my mind. Do you think many jobs will have to be cut?'
'I really don't know at this stage. It often happens, of course, but with good consultants there are ways of making the necessary cuts in the workforce without forcing people to leave. Some will leave of their own accord I expect, when they realise that things will have to change. Some just can't deal with the change. It will be our job to make sure things go so smoothly that they will welcome the new systems.'
Jim wanted to ask lots more questions but the next firm of consultants were shown in and there was a repeat of the morning's meeting but this time with a rather distinguished looking man with an authoritative air. He discussed the 'deliverables' at length and then gave a list of items of information that he would need. Jim realised that although he was still

on a very steep learning curve, he was definitely enjoying himself. He could see lots of work ahead but it promised to be challenging and, if he performed well, also very rewarding. He did wonder, though, whether it would be possible – in the limited amount of time available – to get all the information together. Certainly at MML the systems wouldn't provide the detail required without a lot of digging.

Points to reflect on

- Recognise the major cost drivers for ERP:
✓ People
✓ Training/Education
✓ Time
✓ Changes in processes or ways of working
✓ Problems – there will always be the need for contingencies
✓ Cost of IT, Technology, software, hardware

- Recognise the possibility of outside help on a major project such as this
- Bringing in outside help, consultants will add to the questions and anxieties already associated with the project
- There will need to be data gathering – even at this early stage – to provide details for the initial scoping of the project
- Don't underestimate the amount of change that is starting. Resistance to change will also begin to increase

Chapter 5 – First view of objectives and deliverables

The following week, as promised, Tony Abbey and John Forsythe came over to MML for the meeting to clarify their joint strategy for the merged company. Jim was not looking forward to it as he felt that Neil Martin would try to hold things back. He had been rather enjoying the freedom he'd found as the main representative of Martin's in the Project so far.

As it turned out, he need not have worried. Tony Abbey was a real leader and was obviously determined to make the merger succeed. He was well aware of the problems at MML and skilfully led Neil into agreeing that the strategic direction should be based on the customers' needs. He referred back to their pre-merger discussions with Neil where they had had extensive discussions about customers' needs and when their key performance requirements had been stated at length. They had included improvements to a number of areas - information flows, the supply chain, customer service and order fulfilment. As Tony had put it then – 'we need the right product; right place; right time; right price and right service.'

'I really feel, Neil, that this is our chance to make the most of our combined resources by improving our joint performance in these vital areas – we can make our products better, faster and cheaper. It's our chance to make a quantum leap forward, to overcome our history of resistance to change. The merger seems to me to be what will force two ordinary companies like ours into change that will see us become a major player in our market.'

Jim felt his jaw drop slightly but tried to remember to make an effort to appear businesslike. He didn't think he'd ever heard anyone speak about MML in that way and he could see that it had had the desired effect on Neil. His chest puffed out, his face had gone red and he was nodding so hard it seemed his head was in danger of rolling across the boardroom table.

Tony continued to develop his view of the key performance requirements and handed round a five-page document showing the projected cost savings and benefits. 'The message I really want to deliver today is that we must – we absolutely must – focus on our customers' requirements.' He paused and looked at the expectant faces. 'And what do our customers need? They need the goods on time and in full and to do this we must set demanding Key Performance Indicators and then we need to attain a 99% success rate. ERP will enable us to see failures. Then we can make improvements – continuous improvement. As you can see, gentlemen,' he said 'if you follow all these improvements through to their logical conclusion, then the benefits to our new company will be enormous. If we can get the right consultants on board to push these changes through and help us with the inevitable problems that it will throw up, then we will have something to celebrate.'

Jim looked up at Neil who, for the first time that Jim could remember, actually looked enthusiastic about possible changes.

Tony started speaking again. 'One last point before we wrap this up. I would like us all to sign a 'Statement of Intent' ' They all looked up at Tony with a mixture of curiosity and wariness. 'I would like us all to commit to actively supporting this change project. We should state the company's objectives, sign it and then publicise it. Is everyone on board with that? It will certainly go a long way towards informing all the staff plus customers and suppliers.' He looked expectantly at the other three men.

Without hesitation John replied 'No problem at all. Great idea.'

Jim nodded enthusiastically while Neil pondered his reply.

Eventually Neil spoke 'Sounds fine – in principle. Who's going to put this document together?'

'Oh, that's no problem. I'll do that.' said Tony 'Jim and John, could I rely on you two to look after the circulation – all the notice boards and so on. I'll put something together to go out to customers and suppliers.'

A couple of weeks later John and Jim met up to go through the proposals received from the consultants.

'It might be useful at this point' said John ' to go through what we've done so far and where we go next.'

'Great.' Jim nodded

'Well, we've acknowledged the need for change – it's obvious that the systems from both our companies need to be updated and there is no way that they could be made to work together without fairly drastic changes.'

'Definitely.' said Jim with feeling. He was ready to take notes.

'Okay. We've found problems in both companies. What's your view of the differences in the two companies?'

Jim felt that he was being given the chance to lay his cards on the table about the MML shortcomings. 'The first thing that strikes me is that much more money has been spent at Abbey over the years. At MML we have a lot of old systems and, I've got to admit, inferior premises. Apart from Neil's office, we don't have a single room that looks good enough to take high-profile visitors. Everywhere else is so untidy.'

'Mmm, I'm glad you brought that up. It's an important point. At Abbey we have a 'clean desk policy' so that all desks must be cleared every evening. It's something I picked up from a consultant many years ago. Apparently, research has shown that if the surroundings are right, then the standard of the work is higher. I would guess that you get a lot of minor errors in work at MML – data input errors and so on.'

'That's right.' said Jim with a surprised look on his face. 'I'd never connected the two things. Thought it was just down to the calibre of individual staff.'

'Having good staff helps, of course, but keeping things tidy and clean is an easy way for any company to improve the quality of work. Clean data is invaluable. That might be an area that we can begin to improve at MML even before ERP implementation. Now, I'm not saying that Abbey is perfect but I've been struck by just how unreliable and difficult to get at your data is at MML. We'll see real improvements there if we can start to implement an ERP programme.'

'I guess it would improve all sorts of things. I bet our delivery times would improve if we knew just how much stock we had.'

'Yes, you can see the advantages of better data, can't you? It's vital that we move this along. The sooner we start, the sooner we'll see some of those advantages. There are a number of distinct changes that every project of this type goes through. We're in the Evaluation Stage now so we're going in the right direction.'

'Yes, I can see we are making progress and I'll certainly take on board your comments about a clean-desk policy. I think it might all be a waste of time though.'

John looked concerned. 'What do you mean?'

Jim decided to get straight to the point. 'There's such a contrast between the two companies and the premises... You know what I mean. I think we might end up moving the whole of the Martin operation over to Abbey. It would make sense but I don't relish the thought of all the redundancies. Everyone is worried about their jobs you know.'

'That's inevitable and there will certainly be some job losses. Some will be compulsory of course, I can't deny that, but there will be some natural wastage and some voluntary redundancies don't forget. It won't all be negative.' John replied.

'Well, to be honest,' said Jim 'I'm worried about my own job. I can't afford to lose my job, I've got two families to support. ' He managed a smile but the older man was not convinced.

'Look, Jim, obviously nobody can make any promises in this situation but you've made a good start and anyone who comes full on board with all the changes we're having to make... well, they are more likely to prosper.'

Jim smiled gratefully. 'Thanks John, that's good to know.'

'No problem. Let's go through the proposal from Graham Manders' company.'

Jim could not hide his astonishment at how much the consultants were quoting but John assured him it was in line with market rates and, of course, they wouldn't go ahead if there weren't a real benefit.

'What we really need to get a feel for' said John 'is whether they've got a handle on our two companies and what action they propose next.'

Jim concentrated on the figures and details in front of him in the proposal documents but found it hard to stop his mind wandering to what Neil's undoubted response would be. 'There's just no way that I can see Neil going for this.'

John smiled. 'It's obvious that Neil doesn't like to spend on IT or training but he will have to acknowledge the problems this has caused.'

'I think Tony has already started that process with Neil.' interrupted Jim with a wry smile.

'And also,' continued John 'don't forget that we'll have the consultants' help in demonstrating the benefits. They will give us estimates of what they think can be achieved in terms of savings and increased profits. That will be what will convince Neil in the end. That, and, of course, Tony will definitely add his weight.'

'OK. Let's see what this first one has to say.'

They read without speaking for some minutes then Jim looked up. 'They're asking for so much more information. They say they want to spend three days with us – just to find out our goals and objectives. I'm not sure that I can stand any more information gathering. As you said, it's just so difficult to get to at MML.'

John grinned. 'This is only the start. It will take months. But don't worry; some changes will become so obvious along the way that you'll be able to see the benefits quite early in the project. The real information gathering will be organised by the consultants – that will be the next main stage.'

'I'm not sure that I understand everything they're saying. What's a legacy system for instance?' Jim was beginning to relax with John and felt he could admit his ignorance now.

'Most of it is straightforward. A legacy system is one that we're using right now. It doesn't necessarily mean that we want to get rid of it - we might want to incorporate into the new system. Our stock bar-coding system at Abbey for example.'

Oh, I see' he glanced down at the lists of the consultants' requirements 'I think most of this information can be supplied by our IT and Finance guys anyway – the details and plans for Servers and Networks and the targets for costs and profit.'

'That's right.' said John 'and you should get the Sales Department involved too. It's vital to get people involved. We'll need to understand how all our staff will react to changes – especially the senior guys. Look here,' he pointed to the letter they were examining 'it says they'll run a questionnaire to ascertain readiness for change.'

'That will be fun.' grinned Jim, thinking of Ron and his aversion to paperwork.

'I'm sure you will find it interesting at least. This proposal looks good. The Scoping Workshop is really just for Graham Manders to understand our strategy plus our current processes and systems but it will help us to gather information too. Until that's complete, we won't have to commit ourselves to anything.'

'OK we'll go for that then?' said Jim picking up the next consultant's proposal 'This looks similar, perhaps not quite as professional but not as expensive either. What do you think?'

'I think that the only way we can make the right decision is to see them again. They don't call it a Scoping Workshop but they want to spend the day with some of our senior staff so I guess it amounts to pretty much the same thing. After our next meetings with them, we can get down to comparing their quotations in detail.'

When he arrived home at 7.30 that night, there was no one in the house so Jim sat down quietly for the first time that day and contemplated the similarities between the company merger and the family merger. They certainly had a few legacy systems at home. The girls seemed to have a way of getting into the bathroom first in the mornings and stuck to their routines no matter who was banging on the bathroom door. And Jan still didn't seem to have got used to buying extra soap – she still bought shower gel for the girls and herself – or extra toilet rolls or cereals, any number of the things that they really should be buying in greater quantities now. The girls also stuck to one of Jan's priorities – keeping things tidy. He had to admit that he and the boys were not the tidiest people in the world. That's why the hall was so cluttered that it was like an obstacle course. Perhaps that's the equivalent of untidy desks, thought Jim, we should definitely start a clean hall policy straight away.

'Hi!' came Jan's voice as she came through the front door. 'I've been providing a taxi service for our kids so I've got an hour before I have to go out again to bring them all back from their various activities.'

'I'll go and get them, you sit down.' Jim patted the sofa beside him.

'You sound cheerful. Good day?'

'Yes. It's going well. There's a lot to do, of course, but it's encouraging. I really think we can make this work.'
'Great. What's new then?'
'Well, we've gone through the consultants' proposals and they pretty much make sense to me. I've learned about legacy systems. Your bathroom routine is a legacy system and I'm not sure if it will fit with the new systems we'll have to implement.'
Jan looked up sharply and was relieved to see he was joking. 'I'm sure we'll work it out when we have a new house or an extension. Everyone will be staying in tomorrow evening so we'll be able to go over the drawings for the extension and the rough plans for the new place. Are you sure we're doing the right thing? It's a bit risky with your job isn't it?'
'No, I think the job will be OK. This will be really good experience for me. It isn't just a project for a few weeks you know. The work on this could go on for months and months or even years. We won't be able to put the ERP system in overnight then just leave it. So yes, I do think we're doing the right thing. We need to make our family merger work, it's vital, and I think more space will help so much. I'm looking forward to it, aren't you?'
'Yes, of course I am.' Jan looked at her watch. 'You really need to get a move on to pick up the kids.'
As they all tumbled back into the house an hour later, Jim stopped the boys in their tracks. 'Pick up that Sports Bag and take it to your room, Andrew - and James, hang your coat up. I've decided to implement a clean hall policy!'
The following evening, all six of them sat around the dining table poring over the plans for their extension and also the list of priorities that Jan had prepared.
Jim whistled his admiration. 'This is a pretty comprehensive list, Jan. You're pretty keen on storage aren't you? It's in every part of your list – bedroom storage, hall storage, the kitchen, the bathrooms.'

Jan laughed. 'That will tie in nicely with your clean hall policy, won't it? It's the only way to keep the place tidy. With plenty of storage in the right areas, we can keep everything organised and the place will work for us instead of against us. At the moment, we just don't have anywhere to store some of our things so we're fighting the house – and each other!'

Emma had just spotted the number of bedrooms on the list of suggestions for the new house and let out a whoop of sheer joy. 'Will I be able to have my own TV and friends to stay and a proper dressing table and…'

'One thing at a time, Emma. Yes, you'll get your own bedroom if – and it's a big 'if' at this stage - we go ahead with building a house but you'll have to wait and see about the rest.'

Andrew leaned excitedly over the table. 'There's no swimming pool, though' he wailed

'Put it on the list, Jan.' said Jim 'We should at least think about it.' He gave her a look that said 'Let's humour him. Maybe one day it will be possible.'

'If we build a new house, where will it be?' asked Philippa 'I don't want to move away from my friends.'

'And I don't want to go to a different school.' said James.

Jan held up her hands 'I've seen two plots of land that are up for sale within a few minutes walk of here so if either of them are suitable, then we won't have to change too much.'

'What about the extension plans then?' said Jim

Jan looked thoughtful. 'Well, yes, I'm still keeping the extension in mind but I really think we need to make some quite drastic changes to make it all work so let's see what the architect has to say when we talk to him about this list.'

'You've had a Scoping Workshop all on your own, haven't you?' asked Jim

'What's one of those, Dad?' James sounded interested.

'It's what we have to do at work to look at how all our systems work, how they all fit together and what needs changing. We're going to look at all the company's plans for the future and see if we can do things better. I think Jan has been doing that with us. We all know we need more space but she's had a really good look at where we need the space – what we'll use it for and so on.'

'That's right.' said Jan in surprise. 'A lot of the things you're doing at work will come in handy at home too won't they. That's where the clean hall policy came from.'

'And the need to decide how much soap and so on we need to buy. We need a better forecasting system. All those things need to be worked into our new life in our new home – wherever and whenever that is.'

Points to reflect on

- Strong and visible sponsorship is vital to the change programme
- Data must be recognised as a critical issue. This is usually the number one tactical problem and you must ensure that the data is clean; that you can rely on it when making decisions
- Involve all departments at the early stages of the project
- There are improvements that can be made early in a project that don't require software. This 'low hanging fruit' helps build positive momentum for the programme
- ✓ SKU rationalisation
- ✓ Customer rationalisation
- ✓ Optimise supply chain structure
- ✓ In a merger, rationalise order management

Chapter 6 – Where are we now?

'The Scoping Workshops are arranged for next week' said Jim as he sat down to a breakfast in his pyjamas – it was Saturday so he wasn't in the usual rush for the bathroom!
Jan put some toast on a plate and sat down beside him. 'We should have one here too. I think it's about time we made a decision about the house.'
The following day the rain streamed down the windows and all the children were trying to occupy themselves indoors but the boys were arguing over the computer – again – and the girls insisted there was nothing to do and they were bored.
The Heswall's 'Scoping Workshop' started as an informal chat but as they all became more involved, Jan started taking notes and building up her picture of the family's values and what they needed now and in the future. Seeing the chat develop into a meeting, Jim got out his copy of the quotation from the consultants, which gave the format for the Scoping Workshop.
'Is this what it's really like at work, Dad?' said James 'This is fun.'
'Better than school anyway.' agreed Andrew.
Jan produced her blue book with all the details she'd collected so far and Jim put the quotes from the builders on the table.
'Jan, what's your opinion? Is it an extension – you can see the sort of cost we're looking at – or a new house altogether?'
'Do you really think getting a house built is feasible?'
'Depends on the cost, of course, but I certainly think that we should look into it seriously. Reminds me of something Tony Abbey said. 'I think it was something like 'this is our chance to make a quantum leap forwards' - as a family and where we live.'
'What do you mean?'

'Well,' said Jim 'we could get a lot nearer to perfection if we start with a clean slate. First though, we need to understand where we are and what we do at the moment. That's where your book comes in Jan. Let's go through the different areas.'

'What do you mean, bedrooms, kitchen, that sort of thing?'

'Yes, kitchen first, I think.'

'Right, the kitchen is too small. I'd like us to be able to eat in there – maybe with a table for say, four of us for when we're eating on the run and some of us are out. That would take care of quite a few evenings, wouldn't it? Also, I'd like more cupboard space. I don't have room for all the things from both families. A utility room would be handy too.' She wrote as she spoke and they went through each room of the house and each activity the family did and decided on their possible requirements.

James mentioned the computer. 'That comes under Technology on my agenda.' said Jim in an official voice that made the boys giggle. 'What are your objectives regarding Technology, Andrew?'

He stood up and announced, totally seriously, 'We should make more use of Technology here.'

'Yes, two computers, that's what I said.' chipped in James.

'Costs are important too, you know.' said Jim. 'Not just the building costs but the running costs – we need an economical heating system with enough hot water for a family of six people. I suppose a plumber will be able to tell us how to ensure that. Insulation will be important too.'

Emma started to show an interest at this point. 'We absolutely must be environmentally friendly. The insulation and building materials can be specially chosen. We've done that at school. It really matters, doesn't it, Mum?'

Jan jotted that in her book. 'We'll keep that in mind if it's so important to you, Emma. How does everyone else feel about that?'

They continued like this until they had effectively covered their requirements then Jim announced 'OK. We've made the decision to look into getting our own house built. Unless the costs are prohibitive or something else interferes then that's our preferred route. This won't be easy you know. There will be upsets and hard work. When either of us is busy with this we won't be able to take you out quite so much and you will need to help with the chores too. Are you all ready for the change?' In turn they all replied in suitably sombre voices, that yes, they were ready. Jan knew her girls better than that though and made a note for herself to keep them informed along the way. They were likely to forget what the point of all the hard work was and then there would be problems. Not to mention the sulking that would develop into outright obstructiveness if they weren't careful. She resolved to let Jim in on that – it might help him at work as well as at home.

'What next then, Jan?'

'I'll arrange to see the agents about those plots of land and I'll get Barry and another architect to see us to discuss this. Then we'll take it from there.'

Monday morning Jim whistled as he drove to work. Somehow, even the morning problems with the bathroom were easier to bear now that there was a possible solution in sight.

He spent the day preparing for the workshop and when John called to check on progress and final arrangements, he was amused to hear about Jim's 'trial run' for the workshop. 'Of course, it sounds funny but you will have touched on a lot of important issues there, Jim. Your family's values, understanding how things work at the moment so that you're ready for the 'Data Gathering Stage' and so on.'

Jim was surprised but pleased that John took it so seriously and made a mental note to observe a bit more about the bathroom problem. If they were going to get a new house, it was vital to get it right. It might not be possible to change some things after they had moved in.

The next day the real work started. The three consultants arrived for the three-day workshop and it quickly became obvious that they had carefully worked out their roles and divided the work between them

A lengthy session dealt with strategy. Here John took the floor and although he often deferred to Jim and occasionally to Neville when it came to specifics about MML or about IT and Logistics at Abbey Products, it was evident that he had prepared extremely carefully. First of all he put a convincing case for change and reminded them of what the Project was aiming at and that, as previously agreed, there really was no other alternative. He stated the merged company's problems and opportunities clearly and fielded the questions from the consultants on a variety of matters such as the key issues for the future and the cultural differences between the two companies. By the end of the first day, Jim's head was reeling but he was sure that the consultants – and he – had a good understanding of the strategy on HR, Sales and Marketing, Production and Technology.

The next day the emphasis was on the other members of Jim's team at MML and John brought along several senior managers from Abbey Products. Jim was impressed with the consultants' skills in drawing out information. They asked lots of searching questions about current practices and he found himself a bit embarrassed by his team's lack of understanding of the goals and objectives of the company. It made him realise just how little they had, up to this point at least, been involved in setting the direction and strategy for the company. He found himself reflecting on Neil's management style. Jim knew that on more than one occasion Neil had thwarted his plans for change in his department. It became obvious to him that his difficulties in fitting his job to the company's needs were the result of a lack of direction in the past. Perhaps everyone else had had the same problems. Things were certainly about to change now though.

Just before lunch, a short questionnaire was completed and this was the main topic of conversation over lunch. Again, Jim was in reflective mood and watched and listened as the consultants circulated and chatted with the other executives. He could see that although it all appeared casual, Graham Manders and his two colleagues were systematically gaining their clients' views on issues to be faced. He wondered what they made of Ron who was loudly proclaiming the virtues of the manual systems in the Warehouse at MML. Neville, he noticed, was saying very little, and this gave him cause for concern. Neil, meanwhile, was deep in conversation with Tony and this cheered him a little as he felt sure that Tony Abbey would be using the opportunity to motivate Neil for change.

'How do you think this change will affect your old department, Jim?' asked Graham as he appeared at Jim's elbow with a freshly loaded plate from the buffet.

'I think it can only be good news.' said Jim, putting his own food aside. 'We're long overdue for some re-organisation at MML and some investment in IT. There were one or two goals that I had set for myself in Purchasing that I had not really been able to meet without other departments' help, so I'm looking forward to seeing all the departments drawn together in the next year or so.'

'And you're happy with your own change of direction?' Jim's first impulse was to give an upbeat reply but decided honesty might be the best policy. 'Well, I'm looking forward to the challenge of course, but I'm also a bit apprehensive.'

'That's understandable – change is always unsettling, but what specifically worries you?' said Graham with a small smile, and Jim was struck again by his seemingly casual approach masking a steely determination to get at the truth.

Jim laughed, embarrassed. 'To be honest, taking on the new job worries me. I can do without the lack of job security right now – I've just remarried. That and the worry of whether I can actually cope with everything that this change will throw at me.'

Graham considered his reply. 'Of course, your future prospects could be tied to the success of the project. I know you're just beginning the learning process involved here, Jim, but you've been chosen for a reason. A lot of people have confidence in your ability – me included. Don't forget that we'll give you all the help you need.'

Phew! thought Jim.

Straight after lunch more members of staff were invited into the meeting with the consultants and the consultants were again fact-finding about the systems used in the two companies. This time they carried out an exercise with the staff that they called 'show and tell' which started a bit slowly but soon built into a real discussion as Graham and his team pulled details about the processes and systems from the assembled managers.

Individual meetings with key personnel followed on the morning of the third and final day of the workshop and each of the members of Jim and John's teams was questioned closely on the data they had originally provided.

David Hardcastle, the Production Manager at MML was first up.

'As you know, David, your department is likely to be affected quite a bit by the changes. Are there any parts that you know are not working at the moment? Anything you'd really like to change?' asked Graham.

David laughed. 'Just about everything, I think. The current IT systems don't allow proper tracking of work in progress, we can't allocate specific costs to specific jobs. It's impossible to say which orders are profitable and which are not. The sales department just push orders through regardless of what else will suffer if we change production schedules to accommodate a 'rush job'. Some of the quantities on order are totally uneconomical but it's difficult to prove it.' He looked embarrassed. 'I don't want to criticise anyone else. We all try to do the best we can in difficult circumstances but I have to say that the Sales Department are not performing as well as they appear to be. Neil thinks that Sales are the one area we're getting right at MML but I feel sure that we have big problems there.'

Graham looked straight at David. 'I suspect you're right and that's one of the things that we'll be looking at very closely. Can you give me an example?'

'Well, take the order we're working on right now. It's the biggest order we've ever had, so Neil was delighted with it but it's giving me all sorts of headaches. My entire workforce seems to have been taken over by this order and the sales people know this. It doesn't stop them moaning about the level of complaints they're getting from their regular customers though. Our biggest customers can't get their regular requirements quickly enough so they're placing their monthly orders earlier and earlier. My Production Schedule is getting longer and longer and nobody's happy.'

'Have the changes in the Production Schedule affected quality?' asked Graham, making notes as he spoke.

'Yes, yes they have. Well I think that's the cause anyway. Certainly we've had a surge of quality complaints recently and I can't trace any other likely factor.'

'Any other problems giving you cause for concern?'

'Raw material stock has always been a bit 'hit and miss' – it's difficult to forecast what we need – and, of course, this bigger-than-usual order for one product doesn't help that situation. The Sales people might be delighted with the order as it's helped them to exceed their targets but they didn't really expect to get an order that big.'

They continued in this way, with Graham checking on the information provided by the Production Department and getting David to fill in the detail in several areas.

'Sounds like you've a few problems on your plate, David, what about the new expeditor that you've taken on? Is he helping?'

'Not really. I've tried to help him to set up systems to make the job easier but it's impossible. Everyone argues about priorities. And then tracking actual orders through the shop floor is a skill in itself. I've got to admit that the shop floor isn't the tidiest place in the world.'

'Yes, I can see that. Well, David, our chat's been really helpful to me. I hope we'll get the chance to repay the favour by improving life on the shop floor! In the meantime, I must see the rest of the Merger Project Team. Could you let Bill know I'm ready for him now, please.'

Bill strode into the office and shook Graham warmly by the hand. 'I gather you want to go through some of these figures in a bit more detail.' he said, pointing to the pile of files he had brought in with him.

'Yes, but it's not really the figures I need to deal with just now – those are fairly self-explanatory. Thanks for getting the figures together. I imagine it wasn't easy judging by what I've seen and heard so far!'

Bill smiled. 'No, that's right. It's never straightforward to get figures here. That's where I'm hoping an ERP system will really help.'

'It certainly will.' said Graham. 'What I'd like to know just now is what you think are the main problems to be dealt with.'

'What, in my own area or in the organisation in general?' asked Bill

'Both.' said Graham

'OK. In the Accounts Department, I'm particularly concerned with the growing amount of stock we hold, the length of time it takes us to produce any detailed financial reports and, of real concern to me, the average number of debtor days is increasing.'

'I'm aware of the stock details and of course, as we discussed, the amount of time to produce figures. I think we've already caused you one or two problems in that area.' said Graham with a grin 'but we haven't heard much about the debtor days so far. What do you think is the cause?'

'I think the main thing is the number of customer complaints and queries we're receiving.'

'What sort of complaints?'

Bill shifted in his seat. 'Mostly quality issues but we do have quite a few where we've sent the wrong product or we've delivered to the incorrect place. Some errors are down to sheer carelessness when the data is input. The wrong product codes, delivery details and so on but even when, say, Sales people discover the error the correction doesn't seem to work its way back to the accounts department. My credit control people are continually being told that the invoices are wrong and that someone in the organisation has already been informed. It's a demoralising job for them.'

'Yes, I can imagine.' murmured Graham as he turned over a page and continued his rapid note taking. 'Anything else? Jim tells me you're one of the advocates here for ERP.'

Bill brightened. 'Yes, that's right. From what I've heard and read about ERP, it could be the answer to all our prayers here at MML.'

'I wouldn't put it quite like that' said Graham 'but it would certainly give you the solution to a lot of your problems. We can't just install an ERP system and everything will go right but it does give you a chance to really look at what's wrong – and, more importantly, why it's going wrong – and start to make the necessary changes.'
'Yes, I can see that. Do you think it's likely that we'll go ahead with it?'
'I certainly hope so.'
'Oh, before I go, I mustn't forget to mention the costing system. I don't think there's any doubt that it is a long way from accurate. I don't think we even know which product lines are actually profitable. It's no wonder that the Sales force just sell whatever they can to meet their targets and get their bonuses.'
'Yes, we're looking closely at that area. Anyway, thanks Bill that was really helpful. I hope we'll be working closely together in the future.' They both stood and shook hands before Bill left.
Things continued in this way until Graham had worked his way through the two Merger Teams individually, gaining a depressingly consistent picture as he did so.
The final session involved Neil and Jim from MML and Tony and John from Abbey together with the team of consultants.
'Well, gentlemen, I'd like to thank you for your time. It is certainly an interesting project and one which, I am sure, we will all find extremely rewarding.' announced Graham as he kicked off the Review Meeting. 'We have clarified your key performance requirements and agreed the scope of your proposed changes as well as making a preliminary assessment of the management in the areas most likely to be affected by the changes you wish to make.' He cleared his throat. 'I have to tell you that there is a lot of work to do. That's the bad news. The good news is that there are a number of areas where we can make significant improvements.

We've conducted our initial 'change readiness' evaluations and there are one or two problems there but nothing that can't be fixed. That information will help us to tailor our strategy to get over any resistance and to help you to make the changes happen. We've established benchmarks for the improvements to be measured against and we've identified the necessary starting points with regard to your people, your processes and your technology and I think we've now got a thorough understanding of your business strategy. These are just starting points though and what we need to do now is agree the next steps.'

Neil spoke up. 'What we really need to know is how much this will cost us.' Jim lowered his head and groaned inwardly.

Graham continued smoothly. 'What we would like to do, Neil, is put together a proposal which will show you the sort of return you're likely to get on your investment in the change project. There will be investment in time as well as money. I know already from my own experience that we will be able to deliver savings and improvements worth more than the amount of the cost of our involvement in the project. If I didn't truly believe that, then I would be bowing out of the project right now. But I'm not. I want us to be involved in this. We will lay out all our objectives and the deliverables of the entire project together with a detailed breakdown of our rates and how we've arrived at the project costs.'

'That will be fine.' said Tony 'I'll look forward to that.' His tone was business like and decisive. No further comments were forthcoming from Neil.

Points to reflect on

When starting to build the team:

- ✓ Recognise the competencies needed to make the change
- ✓ Utilise internal and external sources as needed
- ✓ Look at knowledge, experience, behaviour and attitudes

- Recognise that keeping people informed – to the point of over-communicating – will benefit the project
- A programme for ERP will be a major investment in both cash and other resources, such as people's time. Be careful not to underestimate these and weigh them against the benefits that are achievable
- When reviewing issues begin to look at the problems from a process perspective rather than a functional or departmental view, which will have biases

Chapter 7 – Set strategy and vision

Jan and Jim were driving to the outskirts of the next town to view the plot of land that was first on their list.

'It took just fifteen minutes.' said Jan, making a note in her blue book as they pulled up by a large, flat piece of land set back from a quiet road.

They got out and walked around the plot, Jim pacing out what he thought was the size of a kitchen or a bedroom – roughly.

'I just can't visualise it,' said Jan as she stared about her looking lost. 'is it big enough?'

Well, the agent says it is but you're right, it's difficult to imagine a house and everything, with us living here.'

'Well, I'm just as interested in the area. Seems pretty quiet, doesn't it?'

'Lots of places are quiet on Saturday morning. Wherever we decide on, we'll have to come back during the rush hour.'

'And at night time. The whole character of a place can be different at night.'

'That's right. There's a lot to think about for sure.'

They got back into the car and Jim moved to turn the car around.

'Wait, if you drive down there a bit and then turn left, doesn't it lead onto the main road near Melanie's house?' asked Jan.

'Yes, of course. That could be handy for babysitting.'

'Well, these things are important. We need to think about all the things that we need from a neighbourhood. Shall we drive around a bit? Check out the distances to a few places – schools, the swimming pool for Andrew, shops and so on?'

'Let's not forget that the children won't always be going to the same schools. They're growing up. Both Philippa and Andrew will be changing schools soon and James won't be too far behind. Then, before we know it, Emma will be off to college. We should check out the parking space wherever we go – they'll all want their own cars soon.'

'Oh, no! Stop!' wailed Jan but Jim was into his stride now.

'And then they'll leave home and there'll be just the two of us. Perhaps we should be looking for a smaller property.'

Jan thumped her husband's arm. 'Be sensible!'

He rubbed his arm. 'It's something to look forward to though, isn't it?'

'We've a few things to deal with first.'

Back home the blue book was brought out of her bag and Jan started to list the distances from the children's school, from the babysitter's house, and from her workplace. 'When we've decided on the right piece of land, we must check the likelihood of getting planning permission.'

'There's a lot to take into account. Have you considered the shops and entertainment?'

Jan looked thoughtful. I haven't put shops on my list. Having a car, it doesn't seem so important but the land we saw this morning could score well on that point. There's that enormous new supermarket on the route I would take home from work to get to the new house, so that's a plus. What do you mean by entertainment?'

'Well, it would be nice if there was a decent pub within walking distance, perhaps a bowling alley or something handy for the kids. You know, that sort of thing.'

'Yes, and a bit of countryside for a walk. What do you think?'

'I think our wish-list is getting out of control.'

Later, he pondered over what he had said about the wish list. The list of things that were wrong at MML seemed to be growing out of control. He realised that both at home and at work it would be necessary to focus on the key issues. He could see how easy it would be to dream about perfection and not actually achieve very much at the end. He made a mental note to have a chat with Graham about that. Graham's company had by now been engaged to carry out the Business Change Programme and were due to commence work next week although Jim had had various telephone conversations with him in the meantime to prepare for the start of the project. Jim found himself getting increasingly excited – and worried – about the 'change programmes' at home and at work.

'Let's start with a Communication Session.' said Graham in an upbeat manner as he strode into the boardroom at Abbey Products on the first day of the project. Jim's Merger Team had driven over to Abbey and for most of them it was their first view of their new colleagues and the premises they occupied. Most of them, but especially Bill and David, had been openly impressed by what they could see. Jim was, of course, eager to point out the effect that the tidy, well-organised surroundings could have on the quality of work. Tony Abbey, Neil Martin, Neville, plus John and his Merger Team from Abbey were also in attendance.

'We all need to understand the strategy and vision for the newly merged company. I know Tony wants to say a few words on this subject, so why don't you kick off, Tony?'

'Thank you, Graham. Nice to see you here and welcome, of course to all of you.' He leaned forward, displaying the enthusiasm he always showed when speaking on this subject. 'My vision for this company is, as always, success. But as we are now I don't think that's possible.' Neil cleared his throat to speak but was silenced as Tony continued. 'I don't think anyone will deny that we do have some problems. Serious problems. We need to see them from an enterprise viewpoint. At the moment, problems are seen – and tackled – in isolation. For example, the Sales Department look at Sales problems and do their best to solve the issue using the best knowledge they have available. The Production Departments do the same in their own little world. We must have a clear understanding between all the different departments of the company so that solutions are arrived at that are for the benefit of the company and its bottom line.' He emphasised the word 'company' and several heads shot up at the sudden, sharp increase in volume. 'We want to involve people and develop people. Neil and I have discussed the future of the company extensively and, where possible, we have got views from senior management. Our strategy for the company includes an unprecedented investment in people and improving ways of working plus the training to go with it. I've put a bit of detail in this draft document about the company's Core Business Objectives and I'd appreciate it if you could read this as soon as possible, but certainly before our next full meeting on Friday. We'll be testing you then!' He smiled as he handed out the bound reports. 'Seriously, we need to make sure that we all understand these objectives. Just how important that understanding is will become clearer in the coming months as we get stuck into the Business Change Programme. I hope I don't need to remind you that making a success of this ERP implementation and the changes that go with it are imperative for our company to thrive. We all need to work towards that success. The best way we can do that is by co-operating as fully as possible with Graham and his team. With that, I'll hand you back to Graham who I am sure will want to fill you in on a few of the details of the initial steps.'

Jim felt like applauding but Ron Whitehouse most certainly did not feel the same way. He interrupted loudly as Graham tried to take over control of the meeting. 'This is typical. More computer systems! That will just mean disruption to my warehouse and that just leads to poorer customer service. Can anyone tell me how disruption to my warehouse will help my customers?' He slammed his hand down on the table in his usual way as if to say 'this is obvious, there is no argument'.

Jim felt his enthusiasm deflating as he listened in dismay to Ron's loud voice and he wished fervently that the thought and energy Ron put into his negative comments and loud-mouthed assertions could be channelled into something that would help for a change.

Jim looked across at Tony, whose face was surprisingly calm as he stood up to answer. 'I can promise you that the disruption will be kept to a minimum but I am equally sure that change is necessary. Perhaps if we listen to what Graham has to say, we might learn how this will be done.' He smiled tightly, first at Ron then at Graham.

Graham took over with an answering smile to Tony as Ron glared around and folded his arms. 'The main purpose of the next few days will really be to get to know you – how you view the present situation, how you feel about the possible changes to your working practices. We also need to really understand the existing technology so that we can plan for the next stage. That will be the Data Gathering Stage when we'll get down to the 'nitty gritty' and I'm sure we'll get to know one another really well during that phase.' Several people muttered at this point but Graham ploughed on. He made eye contact with most of the people around the table, one by one, and said 'Whatever has gone on in the past or whatever we think of the changes to come, we will all be judged by just one thing – the result.' He looked at all the participants as he said the last two words in a quiet, but serious tone of voice. ' My colleagues and I have already established benchmarks to measure improvements against in the distinct areas – HR, Purchasing, Order Management, Planning, Financial and Manufacturing plus Sales and Marketing. Those are the areas that we – Tony, Neil and I plus several of the board members from both companies – have agreed are the ones to be tackled first. We're planning that they will be involved in our initial modules in the ERP project. There will be more detail on this and all the different stages of the project when we all meet up again at our 'Kick-off' Meeting. That, by the way, will take a full day and we are planning a meal in the evening followed by a 'Quiz Night' and don't worry, the subject of the quiz won't be ERP!' he said looking around as several people answered him with a smile. 'I think we can let you go now while Neil, Tony and I discuss those benchmarks.' He looked directly at John, giving him the cue to gather his papers together. 'John, you and your team have volunteered to give Jim's Team a tour of the facilities here, I believe. That will help you to get acquainted.' He nodded as they left the room and then turned to Neil and Tony. 'I think we should grab a coffee and go through those deliverables again, don't you?'

With coffees in hand, they reassembled around the boardroom table.

Tony gave a rueful smile. 'I don't think everyone's quite on board yet, are they Neil?'

Before Neil could reply for himself or for any of his staff, Graham interrupted. 'I think it went quite well, Tony. Of course, it's very early days and before the end of the week we'll have completed the Change Readiness Evaluation. It is usually done by way of an e-mail online survey. That always separates the men from the boys – or those who are change-ready from those who are change-resisters if we're being more politically correct.' he smiled.

Graham caught up with the Merger Teams just as they were finishing their tour of the Abbey facilities. Jim seized the opportunity to ask him about the project's boundaries.

Graham thought carefully before answering. 'Yes, of course it's possible to let the whole thing spread. Then the danger is that the 'key deliverables' are pushed into the background. It's important to stay focused on the main objectives of the Change Project. As you will see, we'll keep coming back to the main issues time and again. We'll continually measure progress against the key deliverables to make sure that we're still on track. Our regular review meetings will take care of that. We will also put in place the S&OP.' - the sales and operation planning process or 'way of working'.'

At Jim's enquiring look, Graham continued 'That's where we will forecast sales; produce production plans; set out inventory levels and so on. I've got some notes on S & OP processes – they will help you to understand the whole thing and what we're looking for. I'll drop the notes in for you tomorrow.'

That evening at home, Jim reflected on Graham's first day on the Business Change Project 'proper'. Yet again he was struck by the similarities between the problems and issues at home and those at work. He realised that one of the factor's leading to a less volatile change at home was the difference in management styles at home and at MML. Neil Martin's style was extremely paternalistic – he genuinely saw himself as a father figure to many of the staff who had been with him for many years. In turn, they had fallen into habits that they were comfortable with. They saw no reason to change. Neil didn't always hear the truth from them of course. They had learned from bitter experience that he had a habit of 'shooting the messenger' so they kept bad news between themselves whenever possible. It was a strong hierarchy, where everyone – even the relative newcomers - generally knew their place. He and Jan, on the other hand, managed the household by consent and so far everyone in the newly enlarged family seemed comfortable with the set up.

'What are you looking so thoughtful about?' asked Jan as she entered the room.

'Well, it's fascinating really, I was thinking about how management styles differ.'

'Oh, absolutely riveting!'
"It is. We're management by consensus here at home but at MML it's hierarchical.'
Oh, I see.' said Jan, still plainly mystified. She laughed as she picked up some books and one of the boys' jackets.
'You can laugh but it will be interesting to see how that affects our merger situations, won't it?'
'Yes, dear.' murmured Jan as she left the room.
Jim was left alone to wonder just how Graham would deal with the issues at MML and whether or not it really was possible to change the company culture.

Points to reflect on

- Ensure that the ERP programme and its objectives are aligned with the strategic goals of the organisation
- Develop early the milestones and measures for the programme's objectives
- Begin to develop the analysis of your organisation's readiness for change. ERP is about change not software
- As always, communicate with the total organisation about progress, issues, objectives and, most importantly, benefits

Chapter 8 – How to help the organisation change – mobilisation and kick-off

The day of the Kick-off meeting dawned bright, sunny and warm and Graham began to wish that he'd picked another day – a cooler one – especially as the whole day, including the quiz in the evening, would be spent indoors. He knew that people would be even more restless in the heat. He had arrived at the venue for the event early to check that the hotel had everything under control and to go through his agenda one last time.
When everyone was assembled in the large conference room, Tony stood up 'Welcome to the Project Mobilisation Meeting, everyone.' He gave a brief but rousing statement of the event's purpose and then handed over to Graham who was to run the day.
'Welcome to Kick-off.' said Graham with as much enthusiasm as he could get into his voice. 'This is where we really get started. Today, I'm going to set out our plans for the next few months. Tony will be giving you all a bit of background on why the project is necessary – and it is absolutely necessary, make no mistake about that – set out our aims and objectives and so on. I'll be getting the ball rolling by letting you know how I work and setting a few ground rules. Before the end of today you will all be asked to make a commitment. You will be asked to sign your name to the standards of behaviour and performance that will be essential to the success of the project. Tony, Neil, Jim and myself,' Graham indicated each of the men in turn as he spoke 'have already committed ourselves and we expect everyone else to get on board and help this to work. So, if you find that anything is not clear today, please let me know. Good communication is essential and two-way – so speak to me!' He looked around but nobody indicated any desire to speak up. 'As we're on the topic of communication, I can just let

you know at this point that Jim Heswall, who as you know is Project Leader, will be keeping you all up to date with what's going on with a newsletter which will be issued at regular intervals.' All eyes swivelled to look at Jim with renewed interest and Jim smiled bravely back. Inside he was panicking slightly, wondering what he had let himself in for when he had agreed the day before to head up the newsletter. Graham paused and looked around. It had not taken long for them to start looking bored, he thought. 'We've a long way to go and a lot to do but don't worry, it won't be all sitting there and listening all day. I'll be setting you a few tasks – some of them will be fun and some a bit more serious and taxing. Let's just hear what Tony has to say about the background to this project.'

Graham sat down and Tony took centre stage for a short speech before handing the meeting back to Graham who immediately brought up the heading 'Milestones' on his presentation screen and said 'These Milestones will keep us on track and show you what to expect at each stage of the project.'

Graham changed the display on the screen to 'Project Teams'. 'The next thing on the agenda is the formation of teams who will be responsible for making progress in each of the distinct areas of the change project. These teams will report in to the monthly Steering Group meetings.' Graham had given careful thought to the order in which he introduced the different areas for nomination and had also familiarised himself with the people who would be likely candidates for each one. First up is the Manufacturing Team. Any volunteers?' Everyone looked nervously from one to the other. 'Come on. One volunteer is worth ten pressed men – or women.'

Tony stepped into the breach. 'David, perhaps you could lead this team. I'll come onto it as the Abbey representative.'

David signalled his acquiescence and two more volunteers came hesitantly forward while Graham indicated one of his consultant colleagues with particular experience in Manufacturing Operations. Graham moved on to the next team. 'Finance anyone?' John raised his hand without hesitation and was soon followed by Bill and another consultant nominated by Graham. The teams were eventually completed. 'OK, folks, I'll circulate a list of all those teams and a few notes about your responsibilities. Can I suggest you schedule get-togethers for your teams within the next week?'

The day progressed, the intensity of the day broken up with coffee and lunch breaks plus Graham's exercises, some of which caused some hilarity while others produced frowns and groans. The day was rounded off with a general knowledge quiz. Graham stood at the door as all the delegates filed back into the conference room after dinner and randomly split them into teams of six. He knew that the quiz would get them talking and act as an icebreaker.

Jim was sitting at his desk the next day gathering together his thoughts and his notes from the Kick-off meeting when his office door swung open and banged into the wall. 'What's this stupid form that someone's e-mailed me?' thundered Ron as he burst in. He threw the printed pieces of paper on Jim's desk, forcing Jim to look up into the very red face of the Logistics Manager.

Jim had half been expecting some sort of reaction to the Change Readiness Survey that had been sent out by e-mail. He'd warned Graham to be ready for an early response from Ron but even he was surprised by the swiftness and volume.

Before Jim could gather his thoughts for a calming but informative answer, Ron had continued his verbal assault. 'You don't seriously think that I've got time to deal with computers and e-mails asking me stupid questions, do you? I've got a Warehouse to run. You

know that I'm needed to organise despatch of urgent orders – and they all seem to be urgent these days.'
He stopped to draw breath and Jim seized his chance.
'Ron, calm down. All of this is designed to help you. There's some serious investment going into this project, so I think we all need to give it a fair chance.'
'A fair chance!' he bellowed. 'I'll give it no chance! No chance at all! It's all about computers, isn't it? And we all know they don't work. They just make more work for everybody – I've proved that in my department. I do all the proper work and poor Alison has to prove that the work has been done. She has to fill in all those reports on the computer. You tell me how that helps me to get orders out of the door!'
'Well, I...' started Jim, blinking in the face of the continuing onslaught.
Ron threw his hands in the air. 'It doesn't help one bit. That's the answer.'
'Ron, calm down.' tried Jim again 'I suggest you go back to the Warehouse, have a cup of coffee then I'll get Graham to come and have a word with you about it.'
'Oh, no! I don't have time for all this. I'm going back to do some work.' With that he stormed back out of Jim's office, swinging the door shut behind him. Jim breathed out heavily and leaned back on his chair before picking up the phone.
'Graham, Jim here. I've just had Ron from Logistics in here sounding off about not having time to fill in the Change Readiness Questionnaire. Wouldn't listen to a word I said. Thinks all the world's problems are caused by computers. Can you head him off at the pass?'
Graham smiled as he replied. 'This was only to be expected. There are always people who can't – or won't – accept any sort of change and this is the only practical way to find out for sure who they're going to be. I'm sure Ron would have been high up your list, wouldn't he?'

Well, yes, but...'

'No problem.' said Graham. 'I'm on my way to the Warehouse now.' He'd been on site for just a few days but had already settled into his temporary office near to Jim's new office, found his way around and got to know all the key players at MML. It was obvious from the start that the more serious changes involved in the project were going to have to take place at MML, so he'd taken up his position here. Meanwhile one of his junior colleagues was starting the procedures over at Abbey Products. Graham had always thought that finding out who was willing to change – and who was not – was always an interesting process. He'd developed the behavioural and attitude questionnaires himself. They were a vital part of the process as they established the guidelines as to how they should proceed.

By the time Graham walked through the large swing doors into the Warehouse, Ron was sitting behind his desk with a cup of coffee and a newspaper which he hurriedly pushed into a drawer as he explained 'Just having a quick break. Been a busy morning, you know. Can I organise a cup for you?'

'No, thanks.' said Graham as he sat down in Ron's office. 'I've just had one. I understand you've got a problem with the Questionnaire.'

'Oh, no, not really. Just seems a waste of time to me. I'm not convinced that we need another computer system.' Ron was considerably quieter and speaking more politely to Graham than he had been just half an hour previously to Jim. His view was that you never knew how much clout some of these consultants had, so he considered it best to give Graham the 'kid glove treatment'.

'How long do you think it will take you to complete the questionnaires, then, Ron?' he asked with a smile.

'About fifteen minutes, I guess. Computers are not my strong point and you're asking for these to be sent back to you using this thing.' He indicated the computer on his desk that, Graham noticed, had not yet been switched on for the day, with a wave of his hand and a grimace. 'I don't like to look at these damned screens all day, prefer a pen and a piece of paper. I like the hands-on approach, you know.'

'I can appreciate that, of course Ron, but it really is a vital bit of the process. We need to know how you feel about the project. Your opinion, you know.'

'Well, I suppose I can give you the benefit of my experience here. The thing that really worries me here is that we've tried new computer systems before and all we've ever got out of them is extra work. We're always being asked to fill in forms too. Never helps.'

'Are there any of the questions that you're particularly uncomfortable about?' asked Graham, putting a red folder on Ron's desk and getting out a blank printout of the questionnaire.

Ron hesitated. Graham had already guessed that Ron hadn't even looked at the questions, much less considered his answers. Ron capitulated with a show of reluctant cooperation 'I'll see what I can do. Get it back to you tomorrow. Will that be OK?'

'Perfectly OK, Ron. Just let me know if there's anything you need any help with. Or, if you want to discuss anything in more depth, you know where I am. Anyway, I know you're a busy man. I'll let you get on. See you later.' With that, Graham stood up and hurried out of Ron's office and headed over towards Jim's office.

'He didn't bite your head off then?' said Jim by way of a greeting.

'No, no. He was a pussycat. Typical case of resistance but he very kindly wants to give us the benefit of his opinion now, of course.'

Graham leaned back in his chair, obviously settling himself in for one of their lengthy, almost daily conferences about how things were going. The two men had quickly established a routine for working together but they were both still feeling their way through the other's attitudes and foibles. Graham kicked off the session with one of his usual, open-ended questions. 'What did you think about the Project Mobilisation Meeting yesterday?' He looked directly at Jim.

Jim paused, his mind working rapidly, considering his opinions and also what Graham was actually trying to find out from him. Sometimes Graham's questions, he thought, were not as straightforward as they appeared. 'I think Tony handled it very well. For a first meeting, he succeeded in being positive, answered a lot of the questions that must have been in many people's minds. He gave the whole project a mighty push and was convincing when he said that the changes were inevitable. It seemed like a good 'kick-off'.'

'Mmm... you're right he gave us a good start, set the ground rules well - the really difficult thing will be making sure that everyone sticks to them of course. I'll be monitoring report submission dates and perhaps you can keep an eye on any problems about meeting attendance. And although Tony was upbeat – as you would expect him to be – he was also realistic. I was thinking more specifically though. How do you feel about the teams allocation? I know all six of my guys are experts in the fields they've been allocated and they will shadow the team leaders from MML and Abbey but do you think we've got the right people as team leaders?'

'Ah, I see what you mean now. Yes, I think, by and large we've done OK. Mind you, there were still some worried faces when we came away from the hotel last night.'

'Yes, of course, any specific worries from anyone?'

'The usual, you know, job security, will my job still be here when I've finished being a team leader, how many job losses will there be and so on.'

'That's inevitable. We need to manage that situation really carefully. There will be job reallocation, retraining and so on as Tony said, but we can't pretend that no one will lose their job. We must make sure the fears don't derail the project. We'll keep close on that one Jim. The survey will help us too.'

'I was meaning to ask you a bit more about that.'

Graham looked up. 'Like what?'

'Well, I'm not sure exactly what you will use them for – and how.'

'We've got quite a bit of analysis work to do on them – maybe you could help us with that – and I'll go through the results with you. It will help us to formulate a strategy to avoid the people-problems and a communication strategy too, of course. Also, we need to find a few more people who have had positive experiences of change. I'm guessing that there will be quite a few of those at Abbey and we will be able to second them for a bit of extra work. Anything else that was a bit confusing, Jim, or other worries?'

Again, Jim was not sure how much to say but thought that by the time this thing was finished, he and Graham would be like an old married couple so he decided to come straight out with his concerns. 'I must confess that I am a bit worried that I've taken on a bit too much with all this.'

Graham looked surprised. 'Obviously Tony doesn't think that you can't cope or he wouldn't have appointed you and John Project Leaders. I'll help all I can, that goes without saying. But is there anything specific?'

'Right now, apart from thinking that I've got too much to do, no, I don't suppose...' He paused. 'I think maybe the newsletter is a step too far. I've never done anything like it. I can see that it's vital to communicate but...'

'Well,' said Graham, thoughtfully 'Get some help. You really should have someone to do a lot of the work – perhaps one person at the Abbey offices and one at MML?'

'Yes, you're right. I've got someone in mind at MML and I'll check out Abbey.' He went off straight away to have a chat with John about a suitable person at Abbey.

By the end of the week almost all the responses to the questionnaires had been returned. Graham and his team were rounding up the last few stragglers before taking them to do some analysis work on during the next week. They would decide how big a problem they had with resistance to change and formulate a plan to overcome it. They would also ensure that they did not allow anyone to ambush the project. They needed to know where the problem areas lay and the willingness to change – or otherwise – of all the key players. Graham pointed out to Jim that the form was designed to ensure that people had put sufficient detail in their responses and that they would know what people's experiences of change had been in the past, the scale of the changes they had met and what their reactions had been. The majority of the responses had been gathered using the form, which was easy to complete either by e-mail or using digital surveys, and for which the consultants had developed a method of scoring. However, Graham had also spent much of the week interviewing many of the staff face to face. Despite the proven effectiveness of the digital dialogue and attitude surveys, he still liked to get a feel for where the main problems lay and, he hoped, to make an early start on resolving the issues that were inevitably raised.

He rang John to bring him up to date on the process at Abbey. 'Only one missing from Abbey. Not bad, but I'm surprised to tell you that it's Neville's response that we're missing.'

'I'll just put my head round his office door and remind him on my way out if you like.' offered John 'He's probably just overlooked it. This is always a busy time of year for IT and Logistics.'

'Fine. Make sure he e-mails it to me then I'll have it to include in my findings next week.'

On his way out of the factory, Jim called in at Ron's office and found him working on his questionnaire. 'Need any help there, Ron?' he asked.

Ron looked up. 'No, no, I'm just finishing off.' As Jim lingered, the older man took his chance to restate his moans. 'Graham seems nice enough. Keen, you know? But he has no idea what he's taken on here. I've seen all this before. He won't change the way we do things at MML. It'll take a lot more than a whiz kid with a few fancy computers to change this set up.'

Already, Jim was regretting his decision to call in with a few words of encouragement and started to back out of the office before Ron worked his way up to full speed with his usual arguments about how things had always worked. Typical of him to have left it until the last minute, he thought.

'Well, glad to see you're getting on with it. Have a good weekend.' he said raising his hand and waving as he managed to get out of the office just as Ron took a deep breath ready to start again.

He made his escape and continued with the real purpose of his visit to the warehouse.

'Alison, glad I've caught you. I've a favour to ask of you.'

'Ooh, that sounds ominous.' she said with a smile. She was a petite woman, smartly dressed despite working in a rather grubby warehouse.

'Don't worry. I think you might even find it interesting. Challenging certainly.'

'Well, put me out of my misery then. What do you want me to do?'

Jim perched on the end of her desk. It was neat and tidy. 'You've heard about the newsletter aimed at getting the word out to all the staff about the business change project?' She nodded. 'I'd like you to help me with it. Would you, please?'

Her face lit up. 'I'd love to! Now that does sound interesting. Would I have to write articles and so on?'
'That's the idea. Do you think you could?'
'I'd like to try. What about surveys about people's opinions?'
'Sounds great. Ideas already. That's just what we want. Listen, Meg Needham – John Forsythe's assistant - is coming over on Friday morning. She's going to do carry on doing most of the clerical work on this plus getting the news together from Abbey. You'll like her. Very efficient looking. Let's all get together in my office at 10 o'clock Friday. In the meantime, keep those ideas coming.'
'I will. I'm looking forward to it.'

As he let himself in his front door, he could see the two boys fighting in the living room. Emma and Jan were in the kitchen and he put his head round the door, saw that they were deep in conversation and left them to it.
'I'm really pleased that we're getting a new house, Mum.'
'Only 'might be', Emma, nothing's definite yet.' said Jan, but secretly pleased that Emma had raised the subject herself. Since the family's 'Scoping Workshop' both girls had been unusually quiet on the matter and Jan had begun to worry that they didn't really want to move house.
'Well, if we do I'll get a bedroom to myself for sure won't I?'
'Yes, of course dear. I think it's likely that both of you girls will get a bedroom. Maybe the boys will have to share for a bit longer though. Do you think they'll mind?'
'No, I think they like sharing actually. They're only little boys after all.'
Jan smiled to herself at Emma's dismissive attitude to the boys as she continued to chop vegetables. She was finding it difficult to realise that her little girls were

growing up. 'What about Philippa? Is she looking forward to it?'

Emma hesitated. 'I wanted to talk to you, Mum, about Philippa. She's not keen on the idea at all. I think she's a bit upset about it.'

Jan raised her head sharply and turned to look at Emma who was peeling potatoes at the kitchen table. 'Upset? Why?'

'I don't know. I can't understand her. She's such a baby about things like this.'

'What do you mean, things like this? We've never moved house before.'

'Well that's just it. She doesn't want anything to change. She likes everything just the way it is. I think she was a bit nervous about you marrying Jim but she's OK with that now that she's got used to it.'

'Oh, I see.' murmured Jan, her mind racing. She would have to have a little talk with Philippa but, for now, perhaps she could enlist Emma's help. 'Do you think you could help me to make all this go a bit more smoothly then, love? You're the eldest so I would say that you're ready for a bit of family responsibility. It's amazing how much older than Philippa you seem. More mature somehow.' Jan could see Emma draw herself up and preen a little, despite trying to appear nonchalant. 'Maybe you could make sure that Philippa knows that we're only trying to improve things. To make life more pleasant and secure for us all as a family. What do you think?' She looked over at Emma who had stopped peeling the potatoes and was obviously giving this suggestion serious consideration.

'I suppose I could try. I'm sure that she would like a bedroom of her own.'

At this point the kitchen door opened and Philippa walked in, closely followed by Jim. Both Emma and Jan swivelled round, picked up their knives and resumed work, hoping that they didn't look guilty.

'What plans are you two hatching?' asked Jim.

'Nothing' declared both Jan and Emma, in unison.

'Any chance of a coffee then?'

The weekend passed with all the family becoming more excited about the new house plans. In their enthusiasm they had even agreed to forego their trip to Disney World which had been tentatively planned for the summer as their first real trip as a family following their parents' wedding. Even the boys could see that a new house would be a large project and would need a lot of work and all their parents' spare cash.

As soon as he arrived at work on Monday morning, Jim and Graham kicked off the week with an hour's review of progress the previous week and an update on action needed. 'Ron's reply to the questionnaire was here on my desk when I arrived – he's written his answers on the printout rather than online – but we can cope with that. It's better than not completing it at all. Must have stayed late on Friday to do it. It reads a bit like War and Peace.' said Graham.

'He's going to be hard work, isn't he?' It was more of a resigned statement than a question from Jim.

'Yes, but no more than we expected.' said Graham, still keeping up his customary cheerfulness.

'What about Abbey then. Everything shipshape there?'

'One or two problems there too. Nowhere's perfect. One guy hasn't even got his reply back to us.'

'Tony won't be pleased then. He made it abundantly clear that there were to be no acceptable excuses for failure to comply with this. He's really behind this whole thing isn't he?'

'Well, of course he is. As you know by now, I think, there isn't any real choice in this. We either change the company and integrate all the systems and processes or we go to the wall.'

'Yes, that's clear. But who hasn't replied?'

'Would you believe, Neville Crosland, the IT and Logistics Director?'

'Neville!'

'Yep, I've spoken with him this morning already. Says he's going to see Tony about it this morning. He was being very tight-lipped about it. Couldn't get a thing out of him.'

'Very strange. I'll see if Tony knows anything.'

He rang Tony as soon as he put the phone down with Graham but Tony was also keeping quiet on the matter. Something's going on, thought Jim.

Points to reflect on

- Change is what ERP is all about. Understand that even with senior management sponsorship you will still have resistance
- If possible, develop change agents in the organisation that can help deliver the leadership needed to get the change resisters on board
- Be prepared for unexpected issues to arise. Plan for unknown problems with flexible contingencies

Chapter 9 – Gathering the Data

Jim had just arrived at work a couple of days later and had called in at Graham's office on the way through to his own. He was still curious as to what had happened about Neville at Abbey Products but everyone there was being evasive about it. As usual, Graham was already at his desk. He obviously didn't have to queue for a shower in the mornings, reflected Jim.
'Morning, Jim. Have you checked your e-mails yet?' asked Graham as soon as Jim walked into his office. 'Neville's resigned.'
'Resigned!' Jim sat down heavily opposite Graham's desk. He couldn't keep the astonishment off his face.
'You had no idea, then?'
'None, I had no idea at all. I didn't think your questionnaire was that bad!' said Jim with a wry smile. 'Well, that's a turn up for the books. Any other news – his reasons, his replacement or anything?'
'Nothing yet.' replied Graham. 'Just the bald announcement that he's decided to pursue other career options. No doubt it will all come out later but for now we've still got a lot of work to do. Neville's resignation probably won't make life any easier for us. When can we set up a schedule for the next stage?'
'Perhaps I ought to get to my office and check out the e-mails etcetera then we can get together. Say half an hour, my office?'
'OK, see you later.'
Jim walked into his office just as the telephone stopped ringing. He logged onto his e-mails and had just read the 'straight to the point' message from Tony Abbey about Neville when the phone rang again. It was John Forsythe.
'I guess you know about Neville by now, eh?'
'Well, only just. Bit of a shock, wasn't it?'
'Pretty much, although I did think he was acting a little strangely. He's gone anyway. So we'll have to pick up the pieces from here.'

'What will happen? Will he be replaced?'
'Not straight away. As of today, I've been put in charge of IT and Logistics but ultimately, we'll have to find someone with the right experience. In case you haven't noticed, you're a bit weak in Logistics at MML too!'
They talked a bit more about how the Change Project was going and Jim promised to ring John later when he'd had his get together with Graham.
A few minutes later Graham walked into the office and they discussed the Logistics management situation at length.
'I think you and I should meet up with John as soon as possible. Things like this can easily derail a project but at least in John we've now got someone in Logistics who is well and truly behind the project. If Neville wasn't really behind the project then it's better that he's gone straightaway rather than lingering and perhaps causing problems. When you speak with John, could you set up a meeting with him over at Abbey for tomorrow?'
'Sure. Have you got to grips with the Change Readiness Evaluation here?'
'Pretty much.' said Graham 'It was fairly straightforward. We're doing the usual fact-finding. You know, identifying people problems, those who have had bad change experiences in the past. Plus, of course, we're looking for people who've had positive experiences. They could prove to be good team members – we can put them into teams where extra help is needed. Here at MML you've got the usual mix of attitudes and experience. About half are ready and willing to change, I would say, while the other half will need a lot of work. A few problem areas. Funnily enough, Logistics is one of them!'
Jim sighed. 'Why does that not surprise me?'
'Apart from the obvious situation with Neville leaving, there are not too many issues to sort out at Abbey. The bulk of the work will have to be done here.'

'Tell me about it' groaned Jim.

'Don't worry about it for now. We'll keep an eye on the whole situation and see how it works out. Don't forget that we've got the Training Workshops scheduled. They will help. What we need to concentrate on now is setting up the Data Gathering timetable. This is a vital stage. It's our main chance to establish the 'as is' and to involve people.'

'What's the 'as is'? Is it as simple as it sounds?' asked Jim.

'Yes and no. The definition of 'as is' is straightforward. It just means that we determine what the current situation is. From that position we can go on to sort out our 'could be' situation – that's our wish list – and then we can lay out the 'to be'.'

'Right!' Jim was beginning to throw off his despondency about the problems and to get back into the current stage.

He was brought back down to earth as Graham continued. 'Yes, it's simple to say what 'as is' means, but it's a different matter to actually get to the point where you know what your 'as is' actually is.

'OK. Let me in on the secret.' said Jim.

'Oh, there's no secret. It's just hard work. A bit mind-numbing really. Painstaking, detailed recording and analysis of everything that happens on a day to day basis.'

'Yes, I understand that. Where do we start?'

Graham looked thoughtful. 'I think we should dive straight in at the deep end.'

With a sinking feeling, Jim knew what he meant. 'With Ron's Department.'

'That's right. Let's put that first on the schedule for close examination. We'll get into the Warehouse and see how everything is being done. All his systems and processes, who's doing what, how long it takes, who he has to liase with and so on. We'll record all the work volumes and the timings so that we can cross check everything with information that is held on the other systems – the computer stock figures, the job costings and so on. I'll get plenty of information from just walking around, observing the way everyone works and I'll tie everything up with information I'm getting from Malcolm in your IT Department. And to answer your question, yes, I'm sure the Warehouse is the best place to start.'

'Great. Can't wait.' said Jim with a marked lack of enthusiasm.

Graham produced a sheet of paper. 'Next Monday, I think. Will you tell Ron or shall I?'

'I think I'll let you have that honour!' said Jim with a grin. All the other departments at MML were added until Graham and his team of consultants had a full schedule of Data Gathering sessions.

'This is just the start, you know. As I said, it can be boring sometimes, but it can also throw up some really interesting developments and insights and we'll run some workshops to liven things up too. This stage will take quite a while and we'll have to pay repeated visits to all the departments but it's absolutely vital that we get this right. Apart from the essential information that we'll gather, it's the start of getting to know how your people actually deal with change. Lots of changes will have to be made after this and we need to really understand how everything is done. Otherwise, we can't make any changes at all.'

'Do you really think that you can make this happen, Graham? The whole thing seems a bit daunting to me.'

Graham laughed. 'For a start, I am not the one who will have to do it. I'm just the Change Agent here. You and your people are the ones who will have to make it happen. But yes, I know it can be done. You'll see.'

Jim put his head in his hands. 'There are so many problem areas. How can we solve all these problems?'

'For sure you won't get everything right first time. By the end of my involvement here, we will have changed what needs to be changed in advance of the installation of the new software, you'll have a new ERP system up and running and you'll have the knowledge to get on with it. What we will not be able to achieve straight away is 'everything'. Some things will just have to wait until after implementation. It's a case of 'not just now, not all at once'.

Jim groaned again but Graham carried on 'The best advice I can give you at this stage is that this is an elephantine task – it's a large project – and the only way to eat an elephant is in small chunks! Managing a large project like this is best done little by little. If we keep going we will get to where we want to go.'

As he looked at Jim's face, Graham realised that he needed to stop giving Jim the realistic viewpoint and help him into the next stage. 'Come on, let me buy you lunch. You're doing just fine. It's a difficult project – these things always are – but it's a long way from impossible. We'll do it. Let's go.'

Later that day, Jim, Jan and all four children arrived at the plot of land that it was, by now, almost certain they would buy. It was the first time that the four younger members of the family had seen the plot.

'It's just a bit of land, Mum.' whined Philippa. With her arms folded and a bored look on her face, she kicked the tufts of overgrown grass with the toe of her shoe. 'There's nothing to see.'

Jan looked at her younger daughter and sighed. 'Philippa, just use your imagination. You don't usually have a problem with that.' She said wryly, thinking of some of her younger daughter's more creative excuses for not doing as she was asked. 'In fact your imagination is sometimes a bit overactive. Can't you visualise a house, a garden?'

'And a swimming pool.' burst in Andrew.

'Now just wait a minute Andrew.' said Jim as he put his arm around the boy's shoulders.

Philippa stalked off across the plot of land. Jan sighed again. This wasn't going exactly to plan. 'Just leave her, Emma.' she said as Emma made to go after Philippa.

Emma looked up at her mother. 'It is a bit boring, isn't it. Why did you want us to come?'

Jim stepped in. 'It's important that you all get involved. Building the new house will take up all our spare time and energy – not to mention all our money - for the next few months or even the next year, so we want you all to take an interest in it. Ah, here comes Mr Magnusson.'

Five pairs of eyes swivelled to watch - Philippa was by now sulking at the far end of the plot - as the smart looking man approached.

Jim turned to the children and explained. 'Mr Magnusson is in charge of surveying the site.'

Jan and Jim followed Robin Magnusson around the plot as he measured and prodded the ground, all the while discussing their plans for the house with him, although they had already had extensive consultations and had gone through the contents of Jan's blue book in detail.

'Have you considered which way your building would face?' enquired Robin. Jan and Jim and both boys turned to him expectantly. 'With this plot you've definitely got a choice. The conventional way would be to have the front of the house facing the road, of course, and the kitchen at the back but that would give you a north facing lawn at the back,' he said as he studied his compass 'and that's not really the best option.' Jan and Jim looked at each other with blank faces. This was not something they had even considered.

'I think that's something we'll have to give some consideration.' said Jan, scribbling in her notebook.

The boys followed closely behind, watching with curiosity as the surveyor used his various instruments to test and measure the lie of the land. Emma eventually joined Philippa and they both wandered about aimlessly.

As soon as the surveyor had left, Jim took Andrew's arm and explained about the swimming pool.

'Listen Andrew, I think this plot of land is plenty big enough for a swimming pool. But we can't do it just now, not all at once. When we've got the foundations built we'll come back and have a look. Just you and me. We can plan the whole thing so that we know what we're aiming at.'

'Come on girls.' shouted Jan as she strolled across the ground towards her daughters who were sitting on a low, broken down wall at the very farthest edge of the plot. They looked up but didn't get up from their perch as their mother approached. Jan sat down beside them. 'What's the problem here, then?'

'There's no problem really. It's just that Philippa's a bit worried, that's all.' said Emma. 'Tell her, Philippa.'

'I'm not worried.' Philippa's bottom lip quivered and Jan put her arm around her.

'Just tell me what's on your mind, love. Is it moving house that's worrying you?'

'I suppose so.'

'Oh, for goodness sake, Philippa.' exploded Emma 'I'll tell her. She's worried that everything's changing. She likes things just as they are. Even though we do have to share a bedroom and we don't have enough room in the house.'

'Is that right?' said Jan quietly.

'Well, I suppose so. I just like where we live now. I've never lived anywhere else.'

'What are you worried about? The new house will be great. You'll have a bedroom of your own. That will be good won't it?'

'Yes...' hesitated Philippa 'but there'll be nobody to play with. I like living near my friends and near school. Will I have to go to a different school? And what about my music classes? Will I be able to carry on with those?'

'Oh, Philippa. We're not moving to another country!' exclaimed Emma. Jim and the boys turned around, surprised by Emma's shout as they watched from the other end of the plot.

Jan stepped in. 'I know you're worried but there's really nothing to worry about. Emma's right really. We're not moving far. Just down the road really. Most things will stay just the same. You'll see. Anyway, let's join the guys. We can have a chat about everything when we get home. Just don't worry about it.' She held Philippa's hand as they walked back towards the boys, with Emma following.

'What's up?' started Jim but he was silenced by a look from Jan.

'Let's have a drive around for a few minutes before we go back so that we can all get our bearings and see what our new neighbourhood has to offer.' said Jan as they prepared to leave.

'Your wish is my command, boss.' said Jim as he drove out into the traffic.

They toured around the area with the boys pointing out the local park and a bowling alley while Emma tried to cheer her sister up, commenting on the shops they passed.

When they got home, Jim and Andrew disappeared to the garage leaving James free to play on their computer unhindered for once. Jan and the girls went into the kitchen.

'May I make a cake?' said Emma. Jan turned and looked at her daughter in surprise and Philippa clamoured to help.

'Go ahead. That will be great.' said Jan, reaching the flour from the top shelf. 'See Philippa, things like this won't change when we move. We'll all just live together as a family. You're happy with that, aren't you?'

'Yes but...' she stopped and looked dangerously close to tears.

Jan guided her over to the breakfast bar and sat down. Emma meanwhile clattered about with the baking and pretended not to be paying any attention.

'Everything's changing and I don't like it.' whispered Philippa as tears started to roll down her still babyish cheeks.

'I can understand how you're feeling, but most things will be staying exactly the same, you know.' Jan waited until her younger daughter's tears had subsided and then said 'Wait here. I'll get my blue book and we can go through a few things. It will show you more about it so that you will understand.' Neither of the girls had shown much interest in Jan's workbook before and Philippa looked far from convinced now that it would be of any use but Jan hurried through to the dining room to get it. 'Here it is. Now, look at my lists. Here, for example.' She pointed at their list of requirements for a new house. 'We've agreed that we need more room – another bathroom, bedroom and so on. Now, who do you think will be using those rooms?'

Philippa looked at her mother blankly. 'All of us, I suppose.'

''Exactly.' exclaimed Jan. 'Our family will stay just the same. You, Emma, Jim, the boys and me. And look at our list of activities. We'll still be going to school and work, for swimming lessons and music lessons. Jim and I will still be taking all of you children to do all these things – just the same as always. The only difference will be that we'll be getting more space to do all these things. That space will help us. It's not very pleasant to have to wait for the bathroom in the mornings, is it? It makes us all impatient and then we end up arguing. It also makes us late for work – and if we lose our jobs then that's when things would change. We need this extra space to improve things for us and to make us all more secure. Everyone's a bit scared about change but sometimes we have to change one or two things so that the important things can stay the same.' She looked at Philippa. 'Do you think that will be OK?'

'I suppose so.' she wiped her face on her sleeve and slipped off the kitchen stool to join her sister.
'You can grease the baking tray if you like.'
'Gee, thanks.' said Philippa grinning at Emma 'Can't I use the mixer instead?'
'Go on, Emma, let her have a turn.' urged Jan.
Out in the garage, Andrew and Jim were busy tidying out the tool chest and discussing the best way to check the layout of the pool that Andrew was so keen on. They eventually came up with a plan and Andrew could hardly wait until the foundations were laid.
'There's a lot of preparation for a big project like this, Andrew, getting all our possessions in order will be a good start. Now that you've helped me to make a start on the garage, when do you think we should check over your toys and so on?'
'What about the weekend?' said the boy with surprising enthusiasm. He was eager to get started now that he knew he had a chance of a pool.
That evening Jan and Jim sat down with Jan's blue book and went through their finances very carefully. With the money from the sale of Jim's home before their marriage and staged financing arranged through their bank, it seemed that the new home was starting to become the way forward. They moved on to some of the costings they had pulled together for a new house.
'This list seems impossible.' exclaimed Jim as he looked at the sheets of paper that Jan had compiled. They included everything she had been able to think of in connection with the project from the cost of the plot of land to the cost of carpets and a removals van.
'It's not just impossible. It's terrifying.' His glum face spoke volumes.
With her usual brisk approach, Jan took the list out of his hands and said 'Don't get disheartened at this stage. This is just the start. While we were with the surveyor today, something else occurred to me.' She added more items to the list. 'There will be search fees to make sure the land is OK. Now, while we're looking

at all this. Let's just have a quick chat about which way the house should face, then we can send our list of requirements to Barry for the plans to be drawn up and the detailed costings done. What do you think?'
'I think a south facing back lawn would be ideal, especially if we are keeping a swimming pool in mind for the future.'
'Sounds good to me. Maybe the kitchen could go at the side of the house to avoid it facing the road, then?'
'Yep. We'll go with that. Let's see what Barry can come up with when we send him our list of requirements.'
At MML the following week the Data Gathering one-to-one sessions started. Graham tackled the Logistics Department himself and spent an hour with Ron and his assistant, Alison, before commencing the detailed work.
Ron was keen to get on with his usual routine and was plainly relieved when Graham suggested that he would spend the bulk of the time he had planned to spend in the department dealing with Alison. Despite his obvious relief, Ron still had not given up his authority over his department without a fight. Graham had had lengthy discussions with Ron and had finally agreed that Alison would work with Graham and Jim on the Project where necessary but that regular updates about her involvement and the time required away from her regular work would be sent to Ron. He privately resolved to keep the updates brief and deliberately vague to avoid antagonising Ron. Right now, he needed to get on with the work with Alison.
'I'm going to have to ask you lots of involved questions, Alison, so I hope you've got plenty of patience. I can already see that you know how things run around here.'
Alison beamed at him. 'I'm sure it will be really interesting, Graham.' In the background Ron raised his eyes to the sky with a look that suggested both

boredom and relief – a look that Graham did not miss.

Points to reflect on

- Data Gathering is essential to make decisions based on facts
- You may not be able to do everything now, but you can and should plan for the future
- People may leave the project or company if they do not agree with the direction – be prepared
- Remember to go back and remind everyone of the project's goals and objectives
- Communication is key. Set definite timelines on when to communicate with the project team and other employees

Chapter 10 – Establishing the 'as is'

'Is this the right place to start?' said a dishevelled Jan as she waited for Jim to pass her a large box from the highest shelf in the garage.
'What do you mean?' he grunted as he passed the box down to her and came down the ladder to stand next to her. He looked around him at the dozens of boxes. 'We have to start somewhere. There are things in these boxes that haven't been unpacked since I moved in here.'
'Well, if you haven't used them for a while, perhaps you should just throw them out.' exclaimed Jan 'You and the boys mustn't need them. Anyway, what I meant was that there must be a more organised way to sort out our possessions. Perhaps we should link everything up to the lists we made when we first started planning the move.'
'Ah,' said Jim, the light of understanding showing on his face. 'that's what Graham was talking about. 'Understanding the business processes and looking for opportunities for improvements. Said it would save us a fortune if we did that properly before the ERP implementation. Not sure exactly what he meant by business processes though. He said all the data gathering we're doing would help us to see how all the processes relate to each other.'
'That's what I mean.' exclaimed Jan. 'Come on, let's go inside, have a drink and do a bit of planning.'
Jim followed her inside. He could see how Graham – and Jan – needed to understand exactly how everything worked together but he wasn't sure how that would help them to see how to change anything.
He picked up his briefcase as they passed through the hall – it was stowed neatly away in the cupboard there, in line with his 'clean hall policy' – and laid his notes out on the dining room table.
As Jan joined him in the dining room with a tray of drinks, he found what he was looking for in the notes.

'Here it is. We need to align strategic goals, people and internal systems and create a process map.'

'OK, move over. Let me have a look.' Jan studied the papers. 'Look, it says here you have to define the concept of each process before you can analyse it and map it. Then, you can improve it. Or simplify it. And all this is before you implement ERP?' she looked sceptical.

'It seems so. I think this is a job for life, don't you?'

'Maybe. I think you should just take it one step at a time. I also think that we should be getting on with our own process mapping.'

'OK, Miss Sharp from the Knife Box,' he grinned 'how do you suggest we do that?'

'I think that we should decide what our 'key processes' are first.' said Jan thoughtfully.'

'Such as?'

'Such as things that keep everything running here – shopping for instance, or getting ready for work and school in the mornings, or running the children to all their activities.'

'Ah, I see now. Shopping would be the Purchasing function, getting the children to school would be Distribution and so on.'

Jan laughed. 'Well, yes if you want to put it like that. So are you going to do some 'Data Gathering' on that? Be my guest!'

A horrified look crept over Jim's usually calm features. 'Hang on, does that mean I have to follow you round the supermarket?'

'Maybe we don't need to go so far.' she said hastily. 'But compiling a shopping list might not be a bad idea. Both sighed with relief as Jim prepared to do just that.

'Thinking about it,' said Jim 'shopping might be a good place to start. Is it a key process do you think?'

Yes, it certainly is. When we run out of something like bread or milk or eggs it definitely causes problems with other functions like preparing meals.'

'Well, there you are then. Each process impacts on others.'

'Yes, I understand what you're getting at now. And that helps me at work too. What I don't understand here is why we keep running out of things. Don't you just need to buy in bigger quantities than you used to?'

Jan gave him a withering look. 'Oh, if only it was so easy! Where would you suggest I put all these great quantities of things?'

'Ah. Err... inventory is another process isn't it?'

' Inventory? asked Jan with a puzzled look.

'Well, you know, quantities of things that we keep in hand, stock planning etc. Call it storage if you like but it includes a lot more than just the actual cupboards and shelves.'

'Storage. Exactly.' said Jan with a broad smile.

Jim changed the subject hurriedly to avoid more triumphal gloating from Jan. 'We need to get our list of requirements off to Barry. Will you include our list of 'key processes?'

'Of course I will. That's what we're basing our list of requirements on.' She gave him another withering look. Yet again he reflected on his luck at her being so organised and also that there was more to all this than met the eye. But at least he was learning exactly how processes could impact on one another.

By the time he was back in the factory the following Monday he was glad to be able to confer with Graham to clarify some of his ideas.

'One of the things I'm getting concerned about, Graham, is how we decide where to start.'

Graham smiled at Jim's worried look. 'That's where all this Data Gathering that I'm doing is vital. The starting point will become fairly obvious when we've completed that stage. I'm finding out in detail how you make the business work now'

'Or how we don't make it work in some areas, eh?' interjected Jim

'Well, yes, but I'm just concentrating on what you currently do so that we can get down to some Process Mapping as soon as possible. When I've gathered the data I'll be able to understand the inputs and outputs and also the interrelationships between your processes.'

Jim was immediately transported back to his dining room. This was what he and Jan had been doing. Jan had collected a lot of data in her blue book and it was helping them to understand the impact that problems in one area could have on another. He brought himself swiftly back to concentrate on Graham's advice.

'We need to acknowledge which processes add value and which don't. When we've separated these processes we can deal with them differently. The ones that add value, of course, we'll make the most of – streamline and improve them – but the non value added processes need to be minimised – perhaps outsourced - and, wherever possible, eliminated.'

'I see.' mused Jim 'and I suppose we can't get to that point without really understanding how it all works.'

'Exactly.' said Graham. 'One other thing I think we should tackle at this stage is sending out our Invitation to Tender. John's coming over this afternoon, isn't he, for our weekly meeting?'

'Yes, he is – at two-thirty. Have you decided which software houses will be sent the ITT document?'

'That's what I need to clarify with you and John. I've got four companies in mind for the software. I'll bring the details with me. Two-thirty you said? See you then.'

Jim managed to get a quick word with John before the meeting to arrange some help for the newsletter preparation. 'I've enlisted Alison's help – you know, she works in our warehouse – but that wasn't straightforward. Ron is just so resistant to letting go of any of his authority. Anyway, have you any suggestions who might be good at this sort of thing at Abbey?'

'I think my assistant Meg would be ideal. She's organised, good at written communication, hardworking. What more do you need?'

Jim brightened. 'She sounds perfect. Will that be OK with you then?'

Of course.' replied John with a smile. 'My name isn't Ron Whitehouse!' They walked, laughing at their little joke, into the meeting with Graham.

Graham looked at John and Jim's faces as they sat down at the boardroom table. 'Just two things I need to bring up. Firstly, the ITT document. We need to agree the final format and who it's going to be sent to. The other thing is that I wanted to schedule our first Data Gathering workshop.'

'OK, the workshop first, I think. Who do you want in on that?' John enquired.

'I think we need as many of people from the steering groups as possible. Maybe miss out Tony Abbey and Neil Martin from this first one. They might inhibit some of the more junior managers. Let's ask a few people who're involved just now. Alison seems keen.'

'She's good in meetings too.' said Jim 'Good idea to get her more involved.'

Graham chuckled. 'I think she's plenty involved already. I've been following her and asking questions for hours on end.'

'Great. Who else?' asked Jim

John looked up. 'I'll bring a couple of guys who're 'under the spotlight' with your colleagues at Abbey just now.'

'Fine. That should be enough for the first one. We can concentrate on IT and Logistics processes. How about Friday next week?'

'Dress down Friday.' commented John 'The approach will be quite light hearted I expect. Sounds good.'

'Yes, we can have a bit of fun if everyone joins in. OK with you, Jim?' he looked across at Jim who gave a quick nod. 'Now, about the Invitation to Tender. Here's the document.'

Both the other men stared at the thick pile of papers as it thudded onto the table. 'It's fairly standard. It includes a lot of technical information, of course. It outlines your processes and strategy for the merged companies and highlights the areas that we've already decided are our priorities for change. That will help the software houses know which modules you're likely to want in the first stages.'

'What will happen when they get that?' Jim nodded towards the enormous file.

'Hopefully, they'll be in touch fairly quickly. That should show them that you're serious about ERP and they will arrange presentations and then quotations.'

'If that's the Invitation to Tender, I imagine that the prices will be a bit of a shock - for Neil in particular - when we get them!' laughed Jim, nervously.

John spoke up. 'I know that Tony has already gone through a few typical figures with him so it shouldn't come as too much of a surprise. I know Tony – he wouldn't have let things get this far without being certain that he was going to take things through to a profitable conclusion.'

Graham looked sharply at the older man and nodded. 'That's right, I'm sure. Anyway, let me just point out one or two areas for consideration.' He saw the daunted looks on the men's faces and hurriedly continued. 'Don't worry, most of it is routine stuff that I've used before. Let's just run quickly through the list of areas for change then it's ready. As you can see,' he said, pointing at the list of contents, 'I've listed the areas for change. I've left in Customer Relationship Management as a possible module requirement but, as we discussed with Tony last week, it's likely that we will have to descope a bit there. There are so many different ways of dealing with customers in the two companies that tackling that at this early stage could derail the entire ERP project. We would just run out of time, I think. Tony's idea is that we get the main parts up and running and then come back to that at a later date. I agree with him.'

'So Customer Relationship Management becomes Phase Two then?' asked Jim. John nodded decisively.

As they finished their examination of the hefty document, Graham announced the software houses for their approval. 'PeopleSoft, SAP, Oracle and J D Edwards.' They discussed his choices briefly and discovered that he had worked with more than one of them. 'They've all got good reputations. We need to get their responses to our ITT and then see how we could work together. It should be quite straightforward.'

'Sounds fine to me' pronounced Jim and John in unison.

'Great. I'll get those sent out and keep you informed. That's all from me for now. Thanks for your help, guys.'

'One more thing. If you can spare a bit of time, I think you might be able to help Bill in our Finance Department. He's compiling data on stock figures and warehousing costs and seems to be struggling a bit.' said Jim.

'What's the problem?' asked Graham.

Jim looked through his notes. 'From what he said, he's having a bit of difficulty. Any ideas? Is there a quick way to learn this stuff?'

Graham looked a bit impatient but John stepped in. 'Has he tried using a spreadsheet to sort all this out?'

'Don't think so.' answered Jim 'but all finance people use them all the time, don't they, and I'm sure a lot of the data is already entered into spreadsheets anyway.'

'There you are then I'll go and see Bill while I'm over here. No problem.'

Shortly afterwards the meeting broke up and all three next met up at the first workshop. Everyone was dressed casually and there was a relaxed atmosphere which Graham kept going as he bounded into the meeting room. 'Hi guys! Are we ready for some serious thinking, now?' he grinned as he said this and there were good natured murmurings all round of 'Take it easy, it's Friday' or 'Steady on there. Serious thinking's difficult.'

A long piece of brown wrapping paper Jim thought had probably been appropriated from the warehouse, had been pinned up around the boardroom and one or two people were eyeing it suspiciously.

''I can see you've spotted my secret weapon!' beamed Graham, indicating the brown paper and the many pads of brightly coloured sticky notes that he was handing out to everyone. 'We're going to brainstorm the delivery process today. Let's just introduce ourselves first. I'm Graham and I'm co-ordinating this business change project. As you know, this will bring MML and Abbey Products together as one company and make sure that we create one profitable company. Let's start at the front with this face that's getting far too familiar to me.' He indicated Jim with a grin and Jim introduced himself, smiling around at plenty of new faces as well as a few from MML that he had known for a long time. They carried on around the room with plenty of wisecracks as they went and then Graham took over again.

'What I want us to try to do today is find out the really vital bits of the process of delivery so that we know what to keep and what to discard. Right, where do you think the Delivery Process starts?'

'With an order.' shouted a young lad who Jim had never seen before.

'Well done! Where does that come from and when do you get it?'

The young man looked blank but Alison spoke up, rather nervously. 'The paperwork comes off the computer when the stock is ready. We get a delivery note, a pick list and...'

'OK hang on there, Alison. You're right that the process in your department starts with the paperwork you get from the computer system but how does the computer know that there's an order, or that it's due for despatch, or that you've got the stock? Is it psychic?'

Alison laughed nervously but spoke up clearly. 'Well, of course someone has to put all those details into the system before anything can happen. I put the stock figures in myself.'

'Great. You've got the idea. All the different parts of the system have to interact with each other.'

'So what is the first step in the process?' shouted a voice from the back of the room.

'That's what we have to decide. I want to isolate the despatch process in the Logistics Department but I wanted to establish the interconnectivity first. Now, you must be more specific for this next stage. Let's start with all the paperwork. It's the first sign the Logistics Department gets that there's an order to be despatched. Is that right Alison?' She nodded. 'OK write that down.' She wrote 'picking note' in her neat, rounded writing on a bright pink sticky note that she gave to Graham. He wrote NVA on it with a thick black marker pen and stuck it at the start of the band of brown paper. There were a few puzzled faces but no one asked about his notation. They continued in this way with Graham maintaining the pace and humour of the workshop but gradually weeding out the 'as is' until the brown paper was almost obliterated by the suggestions. When Alison got up the courage to ask why he was adding to their suggestions, Graham asked her to wait for the big moment when he would reveal all.

'Now, just as important as deciding where to start with this process is deciding where to finish. Where do you think the finishing point for the Despatch Process is? Joe.' He looked over at the sharp young man, a warehouse supervisor at Abbey. 'Care for a try at that?'

'Just guessing – when the customer receives the goods? Or signs the despatch note?'

'Could be either of those.' he paused. 'but do we want the van driver back at base?' Several people laughed and shouted out 'No way' or 'Have you seen the state of our van drivers!' A couple more bits of paper were added to the brown paper.

'Any suggestions for the significance of NVA and VA, then?' quizzed Graham.

'Not Very Applicable' yelled Joe,

'Good answer, but close.' Graham looked around as he laughed but there were no more suggestions forthcoming. 'It's Non Value Added and Value Added. There, mystery solved, but why do we want to know that?'

Alison spoke up again. 'So that we can keep the Value Added bits and get rid of the rest?'

'That's it! As far as possible we need to keep all the value added components of the process and minimise the non value added.' As he witnessed Alison's involvement with her job, Graham was becoming more and more impressed with her. 'That's a good job done, folks. I'll get that put into an 'as is' Process Map. I think from what we've teased out at this workshop I can answer the vital questions – 'How?', 'With what?', 'By whom?', 'How long?' and 'How often?'. I'll get that together, discuss with a few other people to make sure I've covered everything and we'll review that document at our next workshop.' He looked around at the curious faces and hoped he could keep their enthusiasm going. 'We won't try to work out the VA and NVA situation right now, we'll save that for our next get together. Just one thing about value added before you go, just remember that non-value added does not refer to the person doing the task. It is just a statement about that specific element in the process. There will be a lot of positive changes and lots more things to consider before we've finished.' With that, Graham ended the workshop and everyone filtered out of the room in a buzz of conversation. Just what I like to see, thought Graham.

> **Points to reflect on**
>
> - Process mapping should come from the people who carry out the process
> - 'Buy in' to the plan must be created
> - Introduce the concept of Value Added and Non Value Added tasks and activities
> - ERP implementation is one of the best times for an organisation to revise and simplify its current business processes
> - Making processes more efficient and less complex prior to implementing software will save significant time and cost
> - If possible, ask customers for input on current processes

Chapter 11 – Processes

'There's a lot of discontented muttering going on out there.' said Jim as he walked into the office Graham was using at MML.

He looked up from his computer. 'Hi Jim. Muttering? What's that about?'

'Seems there's a bit of a split decision on your workshop last week. Some people really enjoyed it. Alison, for instance, was quite fired up by it. She collared me in the car park on our way out on Friday. She's full of ideas. I think she just needed someone to treat her seriously and talk about the job with her. There are a few others though who are a bit worried. Your notes about Value Added have got to some of them. They're talking about job cuts and so on.'

'Yes, that's to be expected. Sometimes people take it personally. They need to understand that Non Value Added refers to the tasks and activities and not to the people. We've got another workshop booked for this Friday with the same group, I'll beef up the session I've scheduled on our plans to reallocate, retrain and redeploy. That should help. In the meantime, we'll have to spread the word ourselves. You know the sort of thing - it's early days, we're changing a lot of things and we're making the company more profitable so that we can make jobs more secure.'

'OK, no problem. What's that you're working on?' he leaned over Graham's desk and peered at his computer screen. The spreadsheet looked complicated.

'It's Bill's spreadsheet about customer orders. I'm just filtering it a bit and adding a couple of pivot tables so that I can do some analysis of customer order patterns and so on. Look, you can see how just a few customers make up the bulk of your business but you've got literally hundreds of small customers on your books who just place a very small order occasionally. I suspect that your company treats them all the same.'

'And it shouldn't?'
'Definitely not! When I've finalised the lists of customers, we'll look at how we can deal with the uneconomic customers. Perhaps deal with them via distributors or impose a minimum order.'
'What if they don't go for that?'
''The only reason we're in business, Jim, is to make money so if they're not profitable, we just have to make them profitable – or stop dealing with them.'
'There are some big changes coming up, aren't there?' murmured Jim.
'There sure are. Listen, have you time to look through a few things?'
'Why not?'
Graham pulled a file from his desk and unfolded papers to show Jim some complicated-looking diagrams. 'This is the Process Map that I pulled together from our first workshop.'
Jim stared at the large piece of paper filled with orange and yellow shapes.
Graham pointed at the diagram. 'We're back to Value Added and Non Value Added processes. The orange shapes are for NVAs – filing, such as this here – and the yellow ones show VA.'
Jim read the legend on the yellow shape that Graham was pointing to. 'Ah, I see now. The customer is prepared to pay for the delivery journey – the yellow shape – but doesn't expect to pay for that orange one.'
'You've got it.' Graham looked pleased – and so did Jim. 'A Value Added process is one where both the customer and the company benefit.'
'So what's next?' asked Jim with interest.
'This is where it becomes a little more creative. I'll analyse each process map and try to take out as many of the NVAs as possible, automate others or at least make sure that they don't take up too much time in the overall process.'
'I see. And, I suppose, this is where we can begin to make some savings.'

'Of course, if it's something straightforward that we can implement immediately without too many repercussions on other processes, then yes, that's just what we'll do.' agreed Graham.

'That's great.'

Graham gathered up his papers. 'There's plenty of work to do in the meantime. I've got to get Process Maps together for all the different processes. I'm hoping that we'll be able to get the basic details together from the workshops for most of them. That way, people become more involved and it helps them to understand the processes and what we're trying to do. If they've suggested the different parts of the processes, then they're more likely to believe that it's being done correctly. It gets around the feeling that I've seen from people in the past when they think that the consultant – or the senior management – don't know what the staff actually do.'

'Yes,' said Jim 'and it gives us a chance to understand them too, doesn't it?'

'It certainly does. Understanding their insecurities is the perfect example.'

Jim got up. 'You're right. I'll get on with reassuring some of the Despatch people. I'll look forward to your presentation. It's before our next monthly Steering Group, isn't it? It's definitely getting interesting! See you later.'

Jim walked the short distance back to his own office and sat down behind his desk. He seemed to spend as much time in Graham's office as he did in his own these days, he reflected. He was surprised to see John walk into the office 'Hi John. How's it going?'

'Fine, fine. Thought I'd check how things are moving along here. Need any help?'

'Actually, there are a few things I need to go over with you.'

John looked at Jim with interest.

'I've just been talking to Graham about our customer base. He seems to think that we need to treat the customers differently according to how important they are to us. What do you think?'

'Sound sensible to me.' replied John 'Are we getting rid of some of the deadwood then?'

'Seems like it.'

'We need to develop a grading system for the one's we really want to keep happy. What about a preferred customer status. A sort of 'gold customer'?'

'You mean like the credit cards give out gold cards to people who spend more and can afford it?'

'Yes, that's it. Sounds good, doesn't it? It wouldn't cost us much at all to make our best customers feel a bit special.'

'I like it. Let's sort out a few details and I think you should bring that up at the next Steering Group.'

Their impromptu meeting had only just finished when Jim's two assistants on the newsletter project arrived. It's just one meeting after another, thought Jim.

Meg and Alison were followed by a young man from the warehouse who staggered under the weight of two large boxes which Meg and Alison eagerly unpacked as soon as their helper had left. Soon Jim, Meg and Alison were sitting around the meeting room table devouring the first copy of their new project. It had already been proofed and they had known that it was as good as they could make it but it was still a wonderful feeling to actually hold the finished product in their hands.

'Everybody happy with this?' Jim asked. The two women nodded and Alison even had tears in her eyes. 'OK then Meg, you take enough copies with you for your place and for the mailshot to our customers and suppliers that you've got organised. Alison, you're responsible for distribution here and for sending out the e-mailed version. Both OK with that?'

'It will be a pleasure to do it. I'm so proud of this.' pronounced Alison quietly.

'Don't forget that we will all have to be ready for the feedback. There will be comments and, no doubt, a few complaints and moans coming in tomorrow. We'll all keep track of them and we'll meet back here next Wednesday to analyse the response. Let's hope it's positive.' Jim picked up a large bale of the newsletters. 'Come on, Meg, I'll put these into you car for you. We've done enough here for today.'

Jim arrived at the office bright and early the following morning, ready to face the reaction to the publication. The morning passed quickly, with Jim taking many phone calls with lots of observations about the newsletter's usefulness and only one or two negative comments from the usual cynical and unenthusiastic members of staff. He noted them all – good or bad. He also had a steady stream of visitors but did not mind the interruptions. The interest generated and the feedback he got was interesting and, without doubt, useful.

Even Tony paid Jim a quick visit. 'The newsletter's been distributed this morning. What is the feedback on that so far?' he asked.

'Mostly positive. Your piece as Project Sponsor seems to have gone down well. One or two sceptics made comments along the lines of 'Well, he would say that, wouldn't he?' but other than that, I think people have welcomed it as another source of information. Yes, I'm pleased with it.'

Most of Jim's evenings had been taken up for the past week or so with brainstorming sessions with Jan as they listed the things they had to do on the house project. Barry Bryson had been round one evening and spent nearly three hours going through their list of requirements before promising to get plans drawn up and get back to them.

Jan sucked the end of her pencil. 'Right, I think we've gone as far as we can. We're just waiting for Barry's input now. He said he'd get back to us this week. I think I'll ring him to check how he's doing.' She picked up the phone and dialled Barry's number that was stored in her blue book. By the time the project was properly underway she would know that number as well as she knew her own. 'Oh, hi Barry, it's Jan Heswall here. Just thought I'd see where you're up to with the house plans for us.' She paused, listening and Jim waited for his wife to finish. 'Really! That's great. Yes, Saturday morning will be fine. Yes, see you then. Bye for now.'

'I gather he's coming round.' grinned Jim. 'Has he finished the plans?'

'Yes. He's bringing them round. He hasn't costed anything in detail yet but he's got a list of the different stages and the people that he would use if we gave him the job. Says we can make some kind of chart to keep track of everything.'

'Oh no, this sounds more and more like work every day. I'm reaching overload today. Shall we pack up for now?'

'OK. It's your turn to get us a drink anyhow.' said Jan with a smile. Jim plodded through to the kitchen.

Saturday came around and Barry knocked on the Heswall's front door before 9 o'clock. He was greeted by Jan who had been up and dressed for over two hours. Jim meanwhile had had a lie-in and had just made it into the bathroom to shower. Emma opened the door and showed Barry through to the dining room. 'Would you like a cup of coffee, Mr Bryson?' she asked.

'Don't mind if I do, young lady. Thank you.'

Jan smiled at her almost grown up daughter and said 'That's great, Emma, I'll have one too.'

She turned back to Barry 'I thought we'd be better in here, where we can spread out our papers a bit.' said Jan, indicating the dining table. They sat down and chatted for a while, waiting for their coffee and also for Jim who rushed into the room, muttering his apologies as he sat down at the table with his hair still wet from the shower.

'Right, let's get down to business.' Barry put his folder of papers on the table and opened up the large sheet of paper containing the plans for the house. He pointed to the complicated drawing. 'This is the side elevation, as you can see and here are the views from the front and the rear. Now, if we turn to this drawing,' he pulled another large piece of paper from his folder. 'you can see it's been exploded to show you more detail. Then we have...'

'Just a second, Barry.' Jan interrupted, smiling to soften her request 'Can we take these drawings in before we move on to anything else?'

'Sorry, Jan. Didn't mean to rush you.' They all turned to the first drawings. 'Is there anything there that I should give you more detail on?'

'No, I don't think so. It seems funny to be looking at a drawing of where we could be living soon. It looks pretty much as I would have imagined it, though. 'OK, let's have a look in more detail. Carry on, Barry.'

He spread out the next sheet of plans and they pored over it for quite a few minutes, discussing the finishes on the exterior, the size and layout of the rooms and how closely it resembled their original ideas.

'I think we're ready for the sixty-four thousand dollar question now, Barry.' said Jim and both he and Jan looked intently at the architect who grinned at them.

'The cost, right?' grinned Barry

'Right.' said the couple, looking nervously at each other and back to Barry.

He produced more papers from his folder and laid them out in front of Jan and Jim. 'I've included the cost of my preparing these plans, my co-ordinating the whole project, the rough cost of materials (although that will firm up a bit at a later stage), surveyor's fees, hiring builders, plumbers and so on plus a contingency amount.'

Jim stared and gulped but Jan was more businesslike. 'How flexible is this estimate, Barry?'

'At this stage, as I'm sure you can appreciate, it's just an estimate. It could go up or down depending on how much you change your requirements, materials prices on the day, that sort of thing. The sooner we are able to start and the closer you keep to the original specifications, the closer we'll be to that estimated figure.'

'Can you go through things in a bit more detail then, while we've got you here.' said Jim, finding his voice again. 'For instance, exactly what sort of internal doors have you costed into the plans, the tiles in the bathrooms, that sort of thing.'

Barry looked surprised. 'Well, I don't usually go into that depth of detail at this stage. I've used average rates for all the basic raw materials. I think it might be a good idea if you go through these documents thoroughly and then we can get together again. I might be able to be a bit more helpful then.'

'Yes, that's sounds like a good suggestion, Barry. We'll do that. Can we arrange another time now?' he replied.

'What about next Saturday?'

'Fine. It could get to be a regular arrangement.' grinned Jan.

After Barry had gone, they settled down again at the dining room table and were joined by Emma and Philippa. The boys were playing at a friend's house next door. They concentrated particularly on the order of work that Barry was suggesting and tried hard to understand the implications. Jim had learned from Graham how important it was to identify the starting point for each process. The girls, however, were keen to see how big their rooms would be and how the bathroom would be laid out so Jan and Jim abandoned their discussions on their budget and the order of work and gave in to the girls. They found out each room's measurements on the plans and Jan read out the details of their existing rooms from her blue book where she had already recorded an amazing amount of detail. 'So,' she said 'both your rooms will be a similar size to what you've got now but you'll have a room each.'

'Sounds great, Mum. So that means that we'll both have more space – the space where the second bed is now.'
'Well, yes, at least that amount extra.'
'And will we be able to choose the carpet, Mum?' begged Philippa.
'Hold on a bit.' laughed Jan. 'We've a lot of things to consider before we get to that point.'
'Well, I want a blue one.'
'And I think a pale grey one would be really cool.' said Emma.
Jim looked on amused. 'Leave them to it, Jan. Does no harm to dream.'
'I know them well – they'll have us in the carpet shop before we know what's hit us.' said Jan dryly. 'And we haven't even worked out what the house will cost us yet.'
Jim pulled the papers towards him and concentrated on the planned order of work. 'He seems to have been quite thorough, doesn't he?'
Jan glanced up from her figures. 'That remains to be seen. We need to examine these with him next week. I'll make a note of all the questions as we go through.'
They set to work and only stopped when the boys came into the house, shouting that they were hungry.
At work, the workshops continued and Graham and his team moved into all the areas of operations, sifting information, recording all the details of the processes they observed. Many of the workshops produced lots of interesting information about the 'as is' and the 'to be' for each department which Graham incorporated into his process mapping. Jim worked closely with Graham and learned plenty as he went along. The presentation that Jim was looking forward to came around alarmingly quickly. Both Jim and Graham had been working long hours to get the Process Mapping to an advanced stage in preparation for the presentation.

The presentation was in the boardroom at Abbey Products, where there was sufficient room for all the attendees – the two Managing Directors, Tony and Neil, the two Project Team Leaders, John and Jim, the two HR Managers Caroline from MML and her counterpart at Abbey, Malcolm and Bill from MML plus Graham and his team of six consultants.

'Good morning, we're getting to know one another rather well by now, aren't we, so I don't think introductions are necessary at this stage. We'll get straight down to business. Graham has got to a point where he needs to update us on his progress. It's obvious that plenty of progress has been made so over to you Graham to tell us all about it.' announced Tony.

'Thanks, Tony. You're right. We are making progress. A lot of the Process Mapping has been completed and some of the analysis work is underway. I thought that this would be a good opportunity to show you an example of one of the Process Maps and to show you where we can start to make some savings. I know that's where many of you want to concentrate.' He looked up at Neil who studied his pad in front of him and did not meet his glance. He then brought a diagram up on the screen linked to his laptop computer. 'As you can see, this represents the 'Customer to Cash' group of processes currently in operation at MML.' He grinned at Bill 'Look familiar, Bill? For everyone who isn't familiar with the 'Customer to Cash' process, I'll just explain. It's a level one process that covers the operations involved from generation of demand through converting the demand into sales, the delivery processes, invoicing and receiving payment. That is to say everything from getting the customer to getting the cash – Customer to cash.' His explanation was rewarded with several enlightened and relieved faces around the table. We'll have a look at how they can be useful.'

Caroline spoke up. 'I'm sorry Graham, I can't see that very well. This isn't familiar to me, so it might help if we had paper copies.'

Graham quickly handed out the notes he'd prepared, referred them all to the correct page and continued without, he hoped, losing their concentration. 'I think you've all – with the possible exception of Neil - been directly involved in the data gathering process where we examined all your work processes in fine detail.' He looked around at them, acknowledging their involvement. 'These diagrams are the next stage in the process – they represent some of the analysis work that we've done and are the start of the process re-engineering that my company has been hired to carry out. Let's look at some of the Orange shapes.' All eyes looked down at the papers in front of them. 'They're the Non Value Added components of the Customer-to-cash process, so here we've got things like legal requirements, filing, credit control and so on. They are items that a customer will not usually pay for, so we want to eliminate or automate as many of those things as possible. With the new systems we're implementing, filing is an obvious candidate for elimination. We can partly automate the credit control process so that the time spent on that will be reduced but, of course, we can't do very much about the legal requirements. If we turn now to the yellow parts, these are the Value Added parts of the process; they provide benefits to, usually, both the customer and the company. Here again, we want to automate as much as possible.' He looked around and saw a few worried faces. 'One thing I must stress here is that it is the processes that we define as Value Added or Non Value Added not the people. We are concentrating on analysing the processes and stripping out any parts of the process that are not needed. Savings can then – and only then – be made by retraining people to use the new systems, redeploying them in new situations or reallocating duties. We can make sure that we have more and more people carrying out Value Added tasks – that's the way that we will make the company more profitable. Again, I stress that it isn't merely by getting rid of people's jobs but by ensuring that the jobs they are doing are linked to profit. One thing I would say is obvious and that is the hierarchical nature of the management structure. What I call silo-ed. This type of structure actually works against people being able to help the company, and themselves, to become more profitable and secure. Whereas,' continued Graham, 'a structure that is based on people working collaboratively across a process is the one that will ensure everyone is working towards the same goal'.

Tony took over. 'Thanks, Graham. I'm one hundred percent behind all that. We can make real savings and, more to the point, real profits, by concentrating on getting these processes right before we implement any software. Perhaps we can take some time later to look at these processes in more detail. In the meantime, how are we getting on with buying the ERP software, Graham?'

'We've got quotes and specifications in from three out of the four software houses now and I've chased up the fourth one. I'm confident that we'll be in a position to make a choice within the next month.'

Tony smiled. 'Great. Let's have some coffee now, then we can go through these Process Maps. We need to identify and understand the issues we will have.'

Graham sipped his coffee while he worked. He explained the process maps briefly 'The 'to be' process maps are challenging to existing structures and to people's roles and responsibilities. That's why we need to be especially careful to communicate well at all stages. This is where we work out how people will work, with what software and so on. This is where people's fears will surface.'

Jim butted in. 'Fears about their job security you mean?'

'Yes, security but also status. And fear of change in general. You will get people who will insist on their ways of doing things being included. We simply cannot let this type of fear derail our project.'

'Ah. You mean like Ron in Logistics for example? He seems to be insisting on doing things his way.' said Jim with a rueful look.

'I'm sure we can find plenty of examples around the place.' said Neil, sharply 'but let's not get too personal on this.'

'You're right, Neil. We need to concentrate on the principles in this meeting. The detail will be down to us on the shop floor.' He looked around at his fellow consultants and continued 'What we're aiming at is eliminating some of the 'rubber stamping' that goes on. We need to transform the manager's role. Make it more about training and development and continuous improvement than about authority levels and supervision.'

The following Saturday Barry was back at the Heswall's house bright and early and they spent a couple of hours going through the specifications for the new house.

'I'm trying not to change too much, Barry – I'm mindful of what you said about the costs moving in line with the number of changes – but we will only have one chance to get this right.'

'You're right, Jan. But remember that not all changes will increase costs. I think just as important is the need for making sure everything runs smoothly.' agreed Barry. 'Let's have a look at the order of work.' He turned to a large diagram. I've put it in this form so that we can see where things overlap. We don't want to be trying to do the plastering before we've put the roof on for example.'

'That sounds like something that John's wife, Brenda said they had experienced when they were having their house extended.' said Jim.

'That reminds me. We really should invite them over here for a meal. We can see how they are and compare building experiences!'

'Yes but we'll have to be quick about it though.' Jim agreed. 'If we go ahead with this, it looks like we'll be fully occupied for quite some time. I'll check out dates with John on Monday. We've got our monthly Steering Group meeting so I'm going over to Abbey Products.'

'Fine. We should concentrate on this for now.' Both Jan and Jim turned their attention back to Barry's plans and diagrams and spent some time trying to understand the implications of the order of work. By the end of the morning they were fairly certain that they were going to be having a new house built and that Barry was the man for the job.

Points to reflect on

- Identify a process –e.g. Customer to Cash – not departments. A process is the group of activities that provide input and create an output providing value to a customer internally or externally
- ERP is about process; structure and people follow
- It is important to communicate – continually
- It is necessary to challenge the 'old ways' and introduce visions of 'new ways'
- Refining processes is change. Again, be prepared for resistance

Chapter 12 – Detailed Design (Part 1)

The following day at the Heswall's house saw everyone staying inside. The rain fell relentlessly outside and all the children were bored. Jim called everyone together for a meeting.

'OK folks, we're going to do some planning. We've decided to go ahead with the new house. 'The boys let out whoops of excitement and the girls smiled. Jim continued. 'I gather everyone is in favour then. As the Special Projects Manager, I've got an agenda for this meeting. The first item is appointing Teams.'

'What sort of teams, Dad?' asked James 'Will we have to carry bricks and things?'

James jumped in to the conversation. 'Can I have a builder's hat? One of those bright yellow ones?'

Jan and the girls laughed. 'Not that sort of team. I'm hoping that we won't have to do too much of the building work ourselves!' The boys looked disappointed.

'No, what I mean is teams that are responsible for certain areas of the planning – making sure everything goes well. The first one I've got in mind is a Pool Team and you, Andrew, should definitely be on that one.'

Andrew looked so happy that he was in danger of bursting with pleasure but Jan looked surprised. 'I thought we'd decided that wasn't an option?' she looked meaningfully at Jim.

'You're right – for now. But that doesn't stop us making some plans for later – perhaps next summer. Andrew and I have a few things that we'll be able to do once the foundations are laid. Also on the pool team, I would suggest, should be Barry and we may bring someone else in later. Someone who has specialised expertise in Swimming Pools.'

'Oh, I see. That's OK then. What's next?'

'Well, we decided that one of our most important processes was Storage, so I think we should have a team dedicated to that, don't you?'

'Yes, definitely. I'll volunteer for that one. Maybe James could represent the boys and Philippa for the girls. We might ask you to make a guest appearance, Jim, when we get to the garage and so on. And of course, we will need to make sure that Barry Bryson or one of his contractors are on almost every team.' she suggested.

'Fine. What about a Kitchen Committee?' Jim crossed the item off his list.

'I suppose that's down to me again.' said Jan with a resigned look. 'Emma, do you want to join me on that one?' she asked as she made a note in her blue book.

'OK, Mum. What will I have to do though?'

'The kitchen is probably the most important room in the house. We make all the meals there, we eat some of our meals there, it's where the washing machine is, lots of things so it's essential that we get it right. We'll be selecting the kitchen units and equipment, helping to plan the layout, colour schemes, flooring, all sorts of things.'

'I see.' answered Emma looking more interested now. 'What about the storage in the kitchen? Isn't that the Storage Team's responsibility?'

'Good point, Emma.' responded Jim 'Maybe then we could hold joint team meetings when some areas overlap. Right, next up is the Bathrooms Team. I'm very interested in the Bathrooms in our house. This is what started us thinking about moving so this must be very important. James and Emma I think, for this one. OK?' Both children nodded. 'Any suggestions for any other teams we might need?'

'Let's have a Outdoors team.' suggested Andrew. 'I'd like to be on that one.'

'That's great, Andrew. One volunteer is worth ten pressed men.' Andrew looked pleased but a bit puzzled at Jan's little saying.

They set up a few more teams until all the main processes and areas had been covered. Sensing that everyone was becoming a little overwhelmed as the enormity of the job they were taking on began to sink in, Jim progressed to the next item on his agenda. 'Next we need to establish a Steering Group.'

'What do we have to do for that? Drive a car?' said James, as he giggled with his brother.

'No boys, I'm afraid it's not that sort of steering. It just means a team to look after everything. We'll have regular reviews just to make sure everything is going OK. I think Jan and I can be the main members of that team but perhaps all of you can take turns at sitting in on those meetings.'

'Good idea, Jim.' said Jan 'Is that OK with all of you?' She looked around the table at the now-serious faces and was satisfied that they were beginning to take a full part in this. It seemed that the project had the potential for providing a learning experience for the kids and also just might help to make them more responsible along the way.

Jim broke into her thoughts. 'OK, meeting closed. Jan, I think we need to have a Steering Group meeting now. Who wants to sit in on this first meeting?' He was astounded when all four children spoke at once. They all wanted to carry on talking about the house. This was progress! 'Right just one thing here then. We need to set some Milestones.'

'Sounds like more building work – get your hard hats, lads.' laughed Jan, winking at the boys.

'This is target setting. Let's set out some milestones so that we have something to aim at. We'll need to go through it with Barry, that's obvious, just to make sure that it is feasible but we can draw up a rough timetable for ourselves. Let's have a look at your book, Jan, and see what might be sensible things to put as milestones.'

Andrew looked puzzled. 'I still don't understand. What is a milestone?'

Jim became businesslike again. 'Pass me the dictionary from the bookshelf, James. Let's do this properly.'

James leapt up, found the large dictionary and heaved it on to the dining table. 'It's very important to ask questions when you don't understand something.' Jim quickly found the word and read out 'pillar set up on road to mark miles, stage, event in life." He looked up from the book and saw that Andrew still looked blank and confused. 'It's a marker for something. An event in life could be our house being finished and us moving in. Along the way there will be other things that we can aim at – the laying of the foundations or the first brick for example or maybe getting the money sorted out.'

'Or going to buy carpets?' suggested Emma.

Jan shot a knowing look at Jim and said 'Well, yes, that will be one of our later milestones.'

Jim hurried on 'It is important that we do things in the right order. If we don't then things will very quickly start to go wrong. It could affect our budget as well as our timetable.'

They went slowly and thoroughly through Jan's lists and built up a timetable of important events in the development of their house. Jan looked thoughtful as they discussed the timings involved in the actual building phase. 'You know, when it's being built – before the roof goes on – all sorts of things could happen that will make this timetable useless.'

'What do you mean, Mum?' asked Philippa.

'I'm thinking of the weather, really. It could rain a lot or...'

'Or it could snow and then we could all build a snowman.' said young James, getting excited at that prospect.

'Don't be stupid.' said Andrew, scornfully ' She's talking about the builders. They won't be building snowmen, will they, Dad?'

'Well, no, but the weather is something we should consider. If it did snow, then it could definitely hold up the building. That's a good point, Jan. We need to build in some contingency time for bad weather.'

Jan scribbled a note for herself on the plan and said 'Yes, let's plan some extra time in. When we've been through this with Barry, I'll transfer the details to my blue book. Then we can check off each stage as it happens. Right, that's it for today.' With that, all the children went off to find something interesting to do indoors as outside the rain still poured but Jan and Jim remained at the table. 'I thought that went really well.' said Jan as she packed away the things they had used. She looked over at her husband. 'You look thoughtful. Is everything OK?'

'What? Yes, they are really interested now, aren't they? I was just thinking about your suggestion about contingency time. We've got a timetable at work for all the changes, but I really don't know whether or not there's any time added in for things going wrong. I must ask Graham tomorrow. We've got a meeting scheduled. Things are getting interesting at work too.'

Back at work the following morning, Graham, John and Jim sat down to review the process for choosing a software provider. Jim took advantage of the few minutes before they set about the main work of the day to ask Graham about a few of the things he had been thinking about over the weekend. He quickly updated them on the progress they were making with the house and asked 'Two things came up that I think should be particularly relevant here. Firstly, milestones. I know we have a rough schedule of when we expect to implement the various stages of ERP but I'm not sure how detailed a plan you've got for yourself.'

Graham smiled. 'Milestone Planning, of course. You're quite right, Jim. Some people do go into a lot of detail when they're scheduling this type of implementation. Is that the way you want to go?'

Jim looked uncertain. 'What's your opinion, John?'

'I'm with you, Jim. The more detail we've got in our map of how to get where we want to go, the more likely we are to get there successfully. Can we build on what we've already got, Graham?' he asked as he looked at the progress report so far.

'Yes, of course we can. We can spend a bit of time later today, Jim, if you like. We'll expand on the stages that I've already laid out. We might want to work with the software suppliers on it too. When we've selected them, that is.' His voice had a harder edge than usual, his impatience becoming obvious. 'This isn't something that we can keep delaying. Can we get on to that now?'

Jim and John exchanged glances, but said nothing, they looked towards Graham as he laid out the tenders they had received from the software houses. 'Let's do a straight comparison of the tenders and then get our preferences in order. I'd like to be able to talk to Tony Abbey about this before he goes on holiday the week after next.'

'As I understand it, Graham,' said John thoughtfully 'we will also need to finalise our choice of modules for the initial phase of the implementation. Tony told me on Friday that we might have to descope a little.'

Jim looked up in surprise. 'That's a bit worrying, isn't it? We haven't even started yet.' he said.

'No, it's fairly standard, Jim.' reassured John. 'It just means that we can't have everything at once – a bit like your boy and his pool that you were telling me about.'

'OK – that's makes sense. Will you be deciding the module choice with Tony next week?' Graham nodded. 'What do you think will have to go?'

'I'm not sure of the reasons for the descoping – other than the budget, of course – but I would expect us to keep HR and Purchasing as they are the easiest to implement . The Purchasing module can give big savings fairly quickly too.' he mused 'We have to keep the financial module. It's just too important for us to lose and it will give us so much information about the merged companies that we just can't do without. That will definitely carry on. Of the other two modules – Order Management and Manufacturing – my money would be on us dropping Manufacturing.'

The surprise and disappointment showed on Jim's face. 'But we've done so much work! And we are already seeing improvements.'

John looked up. 'That might be just the reason why it will be the one to go. If we can make savings and improvements in efficiency in the short term without paying for the module and the implementation of the manufacturing module, then maybe it can go on the back burner until we've seen the benefits from the other modules.' He smiled at Jim's downcast look. 'And don't forget, Abbey Martin – and your young son – can't have everything at once!'

Graham and Jim looked at one another in surprise. 'Abbey Martin? That's the new company name is it? asked Graham.

'When was this announced? I've heard nothing.' exclaimed Jim.

'I don't think it has been announced at all.' stated John. 'A shiny new nameplate appeared on the front of Abbey House this morning. Come to think of it, I didn't see one outside here as I came in.'

Jim frowned but decided to carry on with the job in hand. 'Right then. Let's go through the final choices for the Software suppliers.'

They went through the Tenders carefully, and Graham recorded their preferences in each of the categories he had on his list.

'OK, thanks guys. We've got our front-runner – and it's not the most expensive of the contenders so that should please Neil.' He grinned at Jim and continued 'I think we know who we want and I can present the case to Tony and Neil early next week. I don't think we'll have any problem. We'll get who we want. If I can get this decision ratified then I'll be able to invite a software company representative to the next Steering Group meeting. That will move things along.'

Three weeks later, Jim was back at Abbey Products for the Monthly Steering Committee meeting. He was interested to see how the meeting would go without Tony – as the Project Sponsor – to drive it forward and to keep everyone 'on message'. As it turned out, he was not given the chance to find out. Shortly before the appointed time, Jim was sitting alone, waiting in the boardroom and was joined soon afterwards by Frank Littleton from the software supplier. Fifteen minutes later, the door opened and Graham rushed in, already muttering his apologies. He stopped short when he saw only Jim and Frank waiting there in the otherwise deserted room. 'What's going on? The Steering Group meeting's today, isn't it? Where is everyone?'

'You tell me. I've been waiting here for nearly half an hour. I would have been bored except that there's so much to discuss with Frank here.'

'This is ridiculous. Let me find out where everyone is.' He punched the numbers impatiently into his mobile phone and spoke abruptly to the receptionist. 'Yes, I know Mr Abbey is away.' He paused, listening. 'So put me through to John Forsythe's assistant. I see. I see.' He hung up. 'Well, the receptionist has had a message from Neil Martin - he's not coming - and John has been called into another meeting. I don't know...' At that point he stopped as a woman who he vaguely recognised hurried into the room.

'I'm so sorry, I've only just been told that I should attend this meeting...'

Jim stood up and held out his hand. 'Hello, Meg. Nice to see you again. Are you deputising for John?'

The neatly dressed young woman looked gratefully at Jim as she sat down. 'Yes, he's been called into an emergency finance meeting so he told me about twenty minutes ago that I should take his place. I'm sorry that I'm not really up to speed on the Business Change Project but I can take notes and get clarification on anything important.'

Jim looked embarrassed as he apologised to Frank Littleton. 'I'm sorry, Frank. This isn't much of a start for you on our Steering Group.' He glanced over at Graham who was sitting, with his arms folded, at the head of the large, polished table. He did not look pleased. 'As you've heard, Tony's away, John's in another meeting...'

Graham stood up and gathered his papers together. 'I think we should postpone this meeting until we can get everyone together – or until everyone can be bothered to turn up.' With that, he turned and left the boardroom, letting the door slam closed behind him.

Meg Needham looked down at the floor, while Jim quickly took control. 'I guess that means the Steering Group's meeting is postponed. Frank, let's not waste the time we've got. Can we stay on here and spend an hour or two going through a few things?'

'Sure.' he answered 'I'm not going anywhere. If I can get a bit of background from you at the same time as helping you out, then I'm sure it will assist me on the technical configuration.'

'Well, if that's all, I must get on.' Meg stood up and held out her hand to Frank. 'Nice to have met you, Frank. If there's anything you need on the finance side here at Abbey, just let me know. I'll try to help.'

'Thanks, Meg. I'll see you later.' Jim smiled as Meg left the room quietly.

He turned back to Frank. 'I just need to add a few details to this Milestone Planning that I've started. Can we go through your schedule?'

They sat and discussed the software implementation schedule for about an hour and then both went back to their businesses.

When Tony Abbey returned to work the following week, there was an e-mail from Graham awaiting him. When he learned of the problems with the Steering

Group, he immediately set up a meeting. He requested – in no uncertain terms – the attendance of all the members of the Steering Committee plus several of the people within both companies who were involved in the Business Change Project.

Jim sought out Graham to find out what was going on. 'I gather Tony wasn't amused by the meeting collapse?' he said with a wry smile.

'I don't suppose he was.' responded Graham, briskly.

'So what will the meeting do for us?'

'I would guess that Tony will want to get everyone motivated. If people think that the meetings don't matter, just because the Project Sponsor is away, then the whole project can be affected. It's a question of attitude.'

Jim looked pensive. 'Yes, I can see that. I hope there's not a serious problem. I thought everything was going quite well.'

'Generally speaking, it is going well but we need to keep people going. Oh, by the way, I'm glad you're here. I could do with some help with the configuration documents'

'Sure, but what are configuration documents? smiled Jim.

Graham looked surprised but proceeded to explain. 'You remember I mentioned 'as is' and 'to be'?' Jim nodded. 'Well, configuration documents are just an expansion of that process.

Later that day, Jim was surprised with the speed with which Tony took up his team's suggestion. Tony rang him as soon as he received Jim's e-mail. 'This sounds good Jim. Just what we need. I was really disappointed with the fiasco with the Steering Group meeting while I was away so I'm interested in anything that will help keep the project on track.'

Jim stood up and paced around his office, excited now with his idea. 'I just wanted to find something that would make us all work together and not just leave everything to one or two people. We need to spread the effort.'

Tony got out his diary. 'You're right. Let's set up a meeting. We'll put your idea on the table and also we'll turn it into a motivation meeting too.' He flicked quickly through the pages of his diary. 'I think a Friday would be best, don't you?'

Jim muttered his agreement.

'The Friday after next then. Could you organise that? I think we need all the Steering Group, obviously, plus people like Malcolm, your IT guy, and what's that old guy in your warehouse called? People like that. I'll leave it to you. You might want to finalise the attendance list with Graham. If you need any help with the arrangements, catering and so on, just let my secretary, Maria, know.'

Jim was reeling at the pace of the conversation and was making rapid notes as Tony continued. 'Now, for the meeting you need to get together your suggestions for members for each team. Don't put yourself on each team. Have them report to you at the monthly Steering Group. Appoint a lead member for each team to keep the focus and to make communication easier. I'd like to see you on the Purchasing Team and the Logistics Team. Is that OK with you?

'Sure. No problem.'

'Great. That's fixed then.' said Tony. Then, after a short pause he continued 'You know, Jim, I'm conscious that I took my eye off the ball for a while there. Obviously, there's a lot going on right now. Merging two companies and turning them around doesn't just happen by magic. It's a lot of hard work but that's no excuse so next time I can't make a meeting – if there is a next time – I intend to appoint someone to deputise for me as Project Sponsor. You and John are the obvious candidates for that role. Perhaps you could take responsibility for rounding people up. Don't accept any excuses for non-attendance, ask a lot of questions and so on. Are you up for that?'

Jim was surprised but pleased and replied straight away. 'If you think it would help the project move forward, then I'm up for it.' he said with as much certainty and enthusiasm as he could muster as he thought how much extra work could be involved. It could ultimately make things easier for him though so he was pleased that Tony seemed to trust him and also that he was so enthusiastic about the idea for setting up work teams. Tony's passion for the project was infectious and he couldn't wait to tell Jan.

Points to reflect on

- Strong sponsorship is essential to the success of new ways of working
- Develop the team(s) that will work on the project. Encourage learning by the teams
- Milestones will help to facilitate the achievement of goals and timelines
- The project team(s) must continually communicate not only between themselves but to the entire organisation and, at times, including vendors and customers
- Look for innovative, elegant, but ultimately simplified, process design

Chapter 13 – Preparation

As the weeks sped by, the family had to make many exciting – and sometimes difficult – decisions about the details of the house and slowly but surely it began to take shape. The ground was cleared and the foundations laid. Soon, the whole family were able to visit the site and finally able to actually envisage, with their varying degrees of imagination, the finished house.

One day, straight after eating and clearing away the mess from their evening meal, they all piled into the car and drove round to their plot of land that had now become a building site.

The boys raced around the outer walls of the house and argued about which were the biggest rooms and what activities they would be able to carry out in each of the downstairs rooms. The girls stood in what they had been informed was to be the kitchen and chattered about the possible layout. Jan and Jim meanwhile, paced out the dimensions of the rooms and wondered whether their decisions on the layout had been the right ones. Over the noise of the boys' arguments, they discussed the plans for the house.

'Will there be enough room for the six of us plus some visitors in the dining room?' asked Jim

'Well,' said Jan 'it's a good bit bigger than our dining room at the moment so I don't see why not. I'm more concerned about the kitchen. We need to make some final decisions very soon about the finishes in the kitchen and bathrooms. Do we go for the less expensive option with plain white tiles or do we want to make more of a design statement?' Jim mumbled something in reply as he started to move slowly towards the rear of the house, where the boys were standing. Jan shrugged her shoulders and continued to look around alone. 'I couldn't hear you' she shouted to his retreating back and then more quietly, to herself, 'but then maybe I wasn't supposed to.'

Jim drew alongside the boys in what would be the main outdoor area for the house. 'Hi guys! What do you think?'

James replied excitedly 'It's great Dad. Can we build a den?' Both boys eyed the builders' materials – bricks, planks, scaffolding and a concrete mixer - that were scattered about the site. It was obviously an enticing prospect.

Jim replied hastily. 'No, definitely not. You must leave all these things exactly where you find them. It could be dangerous.' He strode out into the centre of the space, beckoning the boys to follow him. They trailed behind, looking disappointed that they were not to be allowed the freedom to roam the building site as they wished. Jim turned and looked back at the foundations. 'Come on you two. Let's pace out the pool area.' At this the boys scampered across the building site and followed Jim's directions as he instructed them where to stand. He waited until they were in about the right positions then went back to his car. 'Stand right there boys. We'll do this properly.'

'We haven't moved, Dad' yelled James as Jim plodded back through the mud loaded with a large bag of sand and a tape measure. His boots stuck in the sodden earth and the boys laughed as he exaggerated his steps and stumbled, narrowly avoiding falling flat on his face.

'Can you two measure out the distance between you?' he asked as he handed the tape measure to Andrew. 'That will be where the shallow end of the pool will be – eventually.'

The boys measured out the distance while Jim placed bricks in the mud to mark the spot. Then they set to work with the bag of sand. With the boys' help, Jim dribbled the sand along the ground along the planned perimeter of the pool. Then he looked at the boys. 'What else do we need to plan?'

Andrew hopped from foot to foot and stumbled over his words as he answered Jim 'What about the steps,

Dad? Or a diving board? And somewhere for you and Mum to sit?'

Jim tried to keep his face composed as he realised that this was the first time he had heard his elder son refer to Jan as 'Mum'. He also knew that he should reign in the boy's enthusiasm before his dreams went too far. 'This is a small backyard pool, you know, not an Olympic stadium or a sports and leisure complex.'

'But it's our pool, Dad. It's our pool. It will be the best pool in the world, won't it, Andrew?' shouted James, his eyes shining just as brightly as his brother's.

Jan and the girls strolled over to join them. 'What are you all up to?' said Jan with a fond look at the boys.

'We're planning the pool layout.' grinned Andrew.

'The pool?' Jan shot a look at her husband. 'We haven't even decided on the kitchen layout and you're here wasting time on a pool? That could be this year, next year, sometime, never.' growled Jan.

Jim looked nervously at the boys and then put his arm swiftly around Jan's shoulders. 'Don't worry. We'll sort everything out about the kitchen.' Seeing her stony look, he added hastily 'And the bathrooms.'

Jan looked around and realised that everyone was concentrating on her, waiting for her reaction. She blew out a long, resigned breath. 'Just so long as we all realise that we need to get the house sorted out first.' She paused. 'But I suppose there's no harm in planning a bit ahead.'

Jim sighed with relief. 'That's right. Planning – as you always remind me – is the key. We know we can't build the pool just yet. Don't we boys?' He shot a look at the two boys. James and Andrew nodded obediently, their faces subdued. 'But by planning at this stage, it will mean that we don't do anything that will prevent us from having a pool later. We've one or two issues at work like this. We know we can't have everything but it's important to lay the foundations for the future.'

As he drove to work the following morning, Jim pondered his remark about foundations. There were several areas where it had become obvious that it would not be possible to accomplish everything they had set out to do. The final decision on module choice for the first phase of implementation was to be made today. They were to have a meeting of the Full Steering Group and he knew that everyone had their own agenda. First though, Tony was going to hold a meeting aimed, Jim knew, at re-motivating everyone. Tony had been openly disappointed that the last meeting had collapsed in his absence. Jim assumed that this was his attempt to steer the Steering Group back on course. Jim would be glad to see that task accomplished successfully but was also keen to ensure that the foundations for as many of the extra modules as possible were laid. He knew that it wasn't possible to do everything at once – just like at home – but needed to make sure that everyone agreed on what were the first priorities. The decisions on the remainder would have to be broken down into the 'essential' and the merely 'desirable.'

He pulled into the car park at Abbey Products and wondered, yet again, whether MML would ever look quite so professional as Abbey. Despite being in a similar line of business and now, actually being part of the same company, the contrast in the approaches and corporate images was startling. Although, on reflection, Jim had to admit to himself that the improvements were beginning to filter through to outward appearances at MML. The car park was a little tidier these days and that same approach seemed to be having its effect in the manufacturing areas. He wondered if the accident rate was coming down and made a mental note for himself to check the figures with Caroline.

As if the power of thought had worked its magic, his mobile phone rang. 'Caroline! I was just thinking about you.'

'Really Jim? All good thoughts, I hope.'
'Oh yes, no problems. I was just wondering if you had noticed any difference in the figures for minor accidents on the shop floor recently?'
'You're not looking for improvements already, are you? That seems a bit optimistic!'
'I try to be positive.' laughed Jim.
Caroline paused. 'To be honest, Jim. I haven't really got all the figures up to date but it shouldn't take me too long to sort them out. Can I ring you back this afternoon?'
Jim couldn't hide his disappointment. 'Could you make it before lunch? As a favour for me? We've got a meeting with Tony Abbey this afternoon. It might do us some good to have some positive news from MML for a change.' He heard the impatient sigh at the other end of the line despite the crackling and interference.
'OK. Give me an hour.' She hung up.
He could imagine the frantic scene in the HR Department now. He didn't have much sympathy though. Caroline really should have kept the figures up to date.
At least Caroline kept to her promise to Jim. Within an hour she was back with the news. 'I'm not sure how we've managed this, Jim, but the accident figures are slightly down. That's comparing them with last month as well as comparing them with this time last year.'
'That's great news, Caroline. Thanks for your help. I'll speak with David to get his feedback. I would think the improvement is entirely down to better housekeeping, changes brought about by the work we've all been doing. Let's see what his view is. Thanks again, speak to you soon.'
Jim wasted no time. He rang the Production Manager at MML straight away.
'David, Good morning. Just wanted your view of the shop floor accident figures. Have you seen them?'
'Not recently, no. Caroline hasn't issued them for a while now and I've been too busy with all the

workshops we seem to be running with the consultants. This ERP is taking up an awful lot of time, you know.'

Jim spoke quickly as David paused. He didn't want to go too far down that path. 'I've just spoken with Caroline and the figures are good. We can see an improvement. Well done.'

'Thanks. Things are certainly looking a bit more organised on the shop floor recently. There's a long way to go yet, of course, but I think some of the problems that have shown up while we've been 'data gathering' as Graham calls it, have been quite easy to solve. So we've put a few small changes in place. Things in their right place, doing things right first time, that sort of thing.'

'That's great. We're at the Steering Group meeting here this afternoon. This news about the accident rate will go down well, I'm sure. Thanks for that.'

'No problem. We aim to please.' David's tone was noticeably brighter by the end of the conversation and Jim resolved to put his own item on the agenda – 'Upkeep of reporting systems – and the benefits to be gained.'

Jim and Graham got together with John over lunch and they discussed the forthcoming meeting. 'I don't suppose Tony was amused about the last meeting, John?' commented Jim.

'You two have no need to worry. You were both there weren't you? I'm going to be making the public apology on behalf of all the absentees. Tony's absolutely right. We do need to make the point that it is unacceptable to miss meetings just because something else comes up.'

'Yes, he's asked me to restate the routines we agreed at the mobilisation meeting.'

'Well, let's get on with it.' said Jim as he glanced at his watch and gathered his files together. 'It's almost time.' The three men walked into the boardroom a few minutes before two o'clock but Tony was already there. He was already tapping his pencil on the highly polished table.

'Ah, two more early birds!' he exclaimed, rising to shake hands with them. 'How are things over at your place, Jim?'

In view of John's subdued mood, anticipating Tony's reaction to the group's attitude, Jim felt that now might be an opportune time to drop his good news into the conversation. He briefed Tony on the new accident figures, avoiding reference to the fact that they hadn't been available until he had chased them up.

'That's fantastic! Benefits already and we aren't even anywhere near implementation of any of the modules yet.'

'Well, as I understood it, we were to put in place any simple changes and improvements straight away and that's what David has done.'

'Well, he's certainly done well. Any other improvements in the pipeline?' Tony leaned forward as he questioned Jim.

During this conversation several other members of staff had filtered into the boardroom and Tony called the meeting to order. 'Right folks, I want, first of all, to welcome Frank and to formally introduce him to you all,' he smiled at the smartly dressed young man on the opposite side of the table. 'Frank Littleton. He is representing our software suppliers. He's a very important man so please be good to him.' Frank smiled and nodded around the room and then left the members of the Steering Group to their meeting.

'Now, first item on the agenda is the last Steering Group meeting. I thought we'd set the ground rules at the mobilisation meeting but there's obviously some confusion so Graham will be giving us the benefit of a clarification session but first John wants to say a few words on this subject. John?' he looked across as John walked to the front of the room.

'I just wanted to apologise for missing the meeting – and I'm sure I can speak for all the other non-attendees at the last scheduled meeting – and to add my weight to Tony's. It is not acceptable to miss meetings without prior notice and good reason. We're all busy people but we signed up to this project and to the requirement to do things properly. Let's make sure we keep on track.' He sounded positive yet apologetic at the same time and Jim admired his friend's ability to hit just the right note. John sat down and was replaced at the front of the room by Graham who quickly ran through the routines that everyone had committed to at the beginning of the project – including submitting reports on time, full commitment, not to mention attending all meetings. He displayed the full list taken from his mobilisation presentation. 'Anyone not remember those?' He looked around, not expecting any answer. 'OK, let's stick with it.'

Tony quickly took over again. Next item is the activity meeting we've got scheduled for the end of this month. Don't worry. It won't feel like work. We'll have a brief meeting then a nice lunch, then a few games of ten-pin bowling to foster a bit of team spirit. Just want to motivate the troops, you know. Here's a list of invitees and full details.' He handed round a couple of typed sheets and a colourful brochure about the venue and the event. 'What do you think? John?'

John Forsythe 'Looks fine. I'm sure it will help to keep us on track. What about everyone else?' he looked around the room. Most people, including Jim and Graham, were beginning to see the benefit of the idea. A bit of bonding would do the project no end of good. Those who hadn't attended the last Steering Group meeting caught on to the fact that this was much better than the reproach they had been expecting.

People shuffled out of the boardroom, leaving just Jim and Tony behind. 'How is it going, Jim?'

'Fine, fine. Well...' Jim paused, unsure whether to voice his concerns to Tony.

"I know.' said Tony with a look of genuine concern. 'It's hard work. I hope the ideas we've started today will help. I can sense a fall-off in motivation levels.'

Jim brightened. He had no need to tell Tony – he had seen for himself. 'It just seems that we've only just started but already people are falling by the wayside.'

'I think we're doing well.'

'Yes, but keeping it up is the difficult bit, isn't it?'

'You're right. But so long as we have people like you on board, Jim, we'll be fine. Is everything going OK at home?'

Jim smiled and soon found himself deep in conversation with Tony about the similarities between the family changes Jim was going through and the ERP project. 'It's uncanny. The more we do at work and at home, the more the similarities keep popping up.'

Tony laughed and said 'So long as you don't get the two mixed up. I don't think we can afford to build a new factory on top of paying for the software!'

Points to reflect on

- Strong sponsorship and involvement are essential
- The project team must represent all parts of the organisation, not only their functions
- Ensure that the teams understand the strategies and the need for change
- Challenge the potential new ways of working. Are they simpler, lower cost, customer focused? Do they deliver added value?
- Prepare/plan for change

Chapter 14 - Problems and Input

Barry Bryson, Architect and Builder, walked briskly up the Heswall's drive at 9.30 the next Saturday morning, already late for the update meeting. He was there to give them the news on progress on the new build. Fifteen minutes later they were all seated around the dining table with coffees and, courtesy of Emma who was taking her work on the kitchen team seriously, freshly made blueberry muffins.

'These are brilliant!' exclaimed James with his mouth full, spraying crumbs over the papers spread out on the table.

'For heavens sake, James, behave yourself. Don't talk while you're eating. We've got a very important visitor this morning.' Jan smiled across at Barry.

'Don't worry about that Jan. I've got two boys of my own. Well, they're not boys now. Grown up I'm afraid, but they used to be a real handful!'

Jim threw James and Andrew a warning look and turned to Barry. 'I think you've got some news for us?'

'Yes, just wanted to go through some figures with you and have a look at these milestones that Jan mentioned.'

Both Jim and Jan straightened up, ready to face the figures. 'Go ahead.' said Jim.

'Let's start with the materials costings. Most things seem to be going to plan so the estimates are proving satisfactory so far. Just a slight problem with slippage on bricks. That came out of the blue. Of course it will add a bit more to the overall price. There are a few other things that have changed too. I've put the revised figures in my report. It's all there.' He handed them a sheaf of dog-eared papers. He pointed to the relevant figures but moved swiftly on to the areas where costs were still the same as originally quoted. Jim and Jan did not have chance to speak before Barry moved on yet again. 'Now, about the milestones. Do you know exactly what you're looking for?'

Jan spoke up. 'We've put a few ideas together' she said, reaching for her blue book, 'but we were rather hoping that you would be able to point us in the right direction.' She started to read at random from her list. 'Land bought, foundations laid, roof on, electrics completed, plumbing, plastering and so on. What we're aiming at is a checklist so that we know that everything is being done in the right order and on time. What do you think?'

Barry paused, biting the ragged skin around his too-short, grubby nails as he gave the matter some thought. 'Mmm... good idea. It's a bit difficult to pin it down in such detail at this stage though. All kinds of things can happen to hold things up. The weather and so on.' He shifted uncomfortably in his chair. The morning sun was shining directly in his eyes.

Jim could see that Jan was getting a bit impatient with Barry's prevarication. 'Yes, we realise that we need to add in contingency time. We just want to be sure that everything's going well. Surely a bit of Milestone Planning will help.' She could see that he did not take to the idea.

'It will give you some guidelines, that's true.' mumbled Barry slowly. 'Could I have a copy of your list and I'll add some bits of my own and a rough estimate of when each stage will be due for completion and I'll get back to you.' His voice betrayed his reluctance and lack of commitment.

Neither Jim nor Jan knew what to say to that so they passed the list over to Barry and Jan said 'This is your copy, Barry. When can we expect your response.' He seemed to be oblivious to her frosty tone but Jim had certainly picked up on it.

'Can we say you'll ring us before next Saturday?' He didn't wait for an answer. 'Anyway, we don't want to keep you. Mustn't waste this beautiful morning, eh?' He stood up and turned towards the door. Barry followed slowly, his eyes lingering on the remaining Blueberry Muffins.

Before Jim returned to the dining room, he went to the kitchen and made two more cups of coffee. This gave him time to gather his thoughts before facing his wife who, he knew, would be getting increasingly worried by Barry's attitude.

She launched into her concerns as soon as he crossed the dining room threshold. 'Isn't he supposed to be looking after the planning? Isn't that what we're paying him for?' She put her head in her hands and Jim went around the table to stand behind her. He placed his hands on her shoulders. 'He seemed so helpful and efficient, so full of ideas when he first came around here but now it seems as though he just doesn't care.'

The children had been sitting quietly with their heads bowed. They slipped from their chairs and raced outside as Jim gave them a look and a nod. 'Well, he certainly wasn't helpful this morning. Perhaps he's just having an off day.'

'An off day!' screeched Jan 'This is our new home. We're putting everything on the line for this. We can't afford for him to have an off day.'

They exchanged long, worried looks. Jim voiced the thought that was haunting both of them. 'Perhaps we should just have had an extension. But it's too late to go back now. We're in too deep.'

With the streak of toughness that Jim had come to expect from her, Jan snapped back into organising mode. 'I don't want to go back. I want a lovely new home. I think we've got three options here. One, we tackle him, two, we sack him and do it ourselves, or three; we get rid of him and find someone else to co-ordinate the whole thing. What do you think?'

Jim rubbed his face. 'I certainly don't think we can do this ourselves. There's more to a building project like this than we could ever imagine. We need someone who understands the systems they need to put into place to manage a project like this.'

'Well, I don't think that someone is Barry. He's just not sticking to what we agreed – there are no proper reports or comparisons against the budget we agreed, it's just going to get worse.' she exhaled loudly and momentarily looked discouraged but put her chin up and continued 'I agree with you though that this isn't something we know enough about to be successful. So, it's the third option, right?'

'Right. I'm going to ring John. Didn't he or Brenda mention that they had used an architect or someone to coordinate their extension?'

Jan brightened. 'That's right. Ring them now, before they go out for the day.'

'Hi Brenda. Has he. Well, nice day for a round of golf. Yes, they're all fine. Listen, we're thinking of changing our architect for the new house so would you be able to give me some details of the man you mentioned that you had used?' He scribbled on the notepad beside the phone, said goodbye and hung up. As he finished his phone call, Jan was waiting to use the phone to call the new architect right away. Jim held on to the handset. 'Hold on, Jan. We haven't sorted out the situation with Barry yet. Let's think about this.'

'Perhaps you're right. But it's really worrying me now. The costs are rising, the timescale isn't agreed and I've just lost confidence in him.'

'I can see that. I suppose it wouldn't do any harm to check out this new guy. I'll ring him and see what he can suggest. Let's take this a bit more slowly. We rushed a bit with appointing Barry so let's learn from our mistakes.'

Jan hovered as he made the call, anxious to hear Jim's impression of John's contact. 'Well?'

'Seems OK. As you heard I mentioned planning and he said it's vital.'

'We know that, for heaven's sake.' exploded Jan.

'Don't start taking it out on me!' he shouted 'Just take it easy. We can't do much today apart from what we've already done. Let me just tell you what I found out.'

She breathed deeply. 'Sorry. I'm just worried.'

'This guy said he'd send us his literature. Says it gives plenty of information on how building work can be controlled. We can read that and then decide what to do. No pressure.'

'But how did he sound? Do you think we can trust him?'

'He sounded OK but maybe the devil himself would sound OK on the phone. Let's just see what his brochure says. Oh yes, he also mentioned some professional body or other that might be able to give us some help. Said we could look it up on the Internet.'

'Well, what are we waiting for? That's something we can do now. We don't have to wait until Monday for that.' She bustled off to use the boys' computer while they were out.

When she came back downstairs an hour later, Jan was noticeably calmer. 'There's plenty of stuff on the Internet. I don't know why we didn't think of it before. I've even downloaded a list of things to check out before engaging an architect.'

'Great.' exclaimed Jim, feeling relieved 'That should be useful.'

On Monday morning, Jim was almost glad to escape from the nervous atmosphere at home and get back to work. He was conscious that he needed to get his next newsletter whipped into shape and decided to invite Meg over to MML that week. There were lots of things to talk about – problems and inputs. And he needed to use the newsletter to get feedback and to try to influence the right behaviours. The other thing on his mind was Graham's meeting organised for that afternoon. They were to go through some of the data the consultants were working on before their Cost and Benefits Presentation the following week. John

Forsythe was due over at lunchtime. It was going to be a busy day and Jim soon lost himself in all the projects he had on his desk.

John and Jim decided to go for lunch before the meeting with Graham. They had phoned Graham to ask him along but he had said he had too much preparation to do. Jim was secretly pleased about this as he had plans to ask John for a few more details – and perhaps a bit of advice – about using the architect he had recommended.

'Brenda told me that you're having a few problems with the new house?' John enquired as they settled into their seats in the pub.

'Yes, Jan's not a happy lady just now. I must admit, she's got a point. Barry Bryson doesn't seem to have heard of planning. The costs are going up all the time too.'

'Costs always go up with building projects, Jim.'

'Yes, I realise that, but this is more about our confidence in the guy. He just doesn't seem to have his finger on the pulse.'

John took a long drink of his beer. 'That is definitely more worrying. If he doesn't control everything then you won't have any chance of completing on time or of getting the house built just as you want it. Then costs will just spiral out of control. Imagine if our consultant on this project here at work was even half as lackadaisical! It just would not work.' Jim looked glum. 'So, you seem to have decided to try someone else. This firm - Chancery Architects - are good. They're not the cheapest around but they have a really good reputation. Could be worth it for your peace of mind.'

'And my marriage!' interjected Jim with a grimace.

''That's even more important.' John agreed. 'I think you've too much on your plate at work at the moment to be able to find time, or energy, to get to grips with controlling a building project. Far better to get a professional to do it for you. I think you'll be pleased about the controls this firm will put in place. Regular reports, proper planning. The works.'

'Sounds good to me. Let's order some food then you can tell me all about your experiences with building.'

They walked back to the office without having even touched upon the business change project but feeling refreshed and ready to hear what Graham had to say about his Cost and Benefit Analysis.

They entered the meeting room together to find Graham already waiting for them, surrounded by files, loose papers and fancy-looking charts.

'Hi there. Let me just get these in order then we'll go through a few things.' The two men sat down as Graham shuffled the papers and muttered to himself.

'Right. What we need to go through is the effect of the data combined with the work-study we've been doing. Just to update you and run a few ideas by you. When everything is pulled together like this, it's possible to see exactly where we should be going. Of course, the Data Gathering hasn't finished yet and we're a long way from arriving at the final design of the whole project. In the main though, I've found that the results here validate what we originally thought and we can see the Costs and Benefits clearly at this point.'

Jim leaned forward and grabbed a copy of the file that Graham was holding out. 'I assume that means there are plenty of savings to be made?'

'Yes, just as we thought. One of the major benefits though is bringing the strategy of the merged company into focus. As Tony has said all along, getting the ways of working synchronised is essential – the company just will not survive without the information to manage it properly. Apart from bringing all the software, the processes and, of course, the people of the two companies together, we'll also be able to improve performance in most areas.'

'Such as?' asked John.

'Well, look at this.' He pointed to one of his charts. 'I reckon we can reduce headcount by twenty percent and...' He noticed Jim's pained expression and hastily added 'over a period of time, of course. Natural wastage, redeployment and all that. Don't forget that, here at MML at least, you've got an ageing workforce. If the terms are right, many of them will jump at the chance of early retirement.'

'This might be viewed as good news by Neil and Tony but we must make sure that this is presented in the right way to the workforce. News like this mustn't get out. It would cause mass panic if it did.'

'News like this always gets out.' John declared. 'The trick will be to make sure that it gets out in the right way – when we decide it's the right time.'

Graham sighed. 'John's right, Jim. But don't worry, we'll involve the HR people. Anyway, there's a long way to go yet. That's just a rough prediction. There'll be plenty of redeployment and retraining as well as people retiring and so on. Just let's go through these figures. As I was saying, there are various areas of improvement. We will reduce query-handling time. The benefits from that are limitless. Customer satisfaction will obviously improve, there'll be cost reductions in terms of desk space, staff on-costs, the debtor days will reduce by up to forty percent once queries are handled more efficiently. They are all shown on the Credit Control chart. Look at the Delivery First Time chart too. It shows similar savings.

'Looks good, Graham.' admitted Jim. 'This is exactly what we wanted to achieve isn't it?' He looked first at John then at Graham.

John let Jim's question hang in the air, his mind on other aspects of the project. 'I assume you've also finished the cost figures for this Cost and Benefit analysis. It won't be all good news.'

'That's right. The costs are listed on page twenty-nine. It gives figures for the software – both for the current project and the estimates for the modules we've left for later, the training costs...'

'And your costs...' grinned Jim.
'Graham grinned even more broadly. 'Yes, my costs. And very reasonable they are too!' He paused for a few seconds as they flicked through the report. 'That's just the first draft. I've a few more things to add and refine ready for the final presentation to Neil and Tony – and yourselves of course, but I would appreciate your feedback.'
They did not get a chance to comment at that point because there was a light knock on the door and Tony Abbey walked into the room. 'Hi. Sorry to interrupt.' John and Jim moved their chairs to one side, making space for Tony. 'Thanks but I'm not staying. Just wanted to show you this before we send it off to the printers.' He put some mock-up sheets on the table in front of the three men who all peered at the small print on the shapes of bright red and green outlined. The small oblongs were the size of playing cards. 'We're going to have them laminated and I want us all to use them in any meetings about the Business Change Project.'
While Tony had been speaking, Jim had been reading rapidly through the lists on the sheet. Understanding showed on his face.
'What do you think, Jim? Will this help?' asked Tony
'I think so,' answered Jim slowly. 'They're lists of 'dos and don'ts' aren't they?'
'That's a good way of putting it. We must foster more positive ideas and behaviours in these meetings. I read about this idea in a book. Thought it would help us a lot. But, make no mistake, if people repeatedly display 'Red Card Behaviours', action will be taken - performance reviews and so on.'
'I'm sure we're all grateful for any help we can get, Tony.' said Graham.
'Good.' pronounced Tony. 'I'm going to launch the Red Card/Green Card system at the Activity Day but it will be mainly up to you three to make it work. I'll leave those proof copies with you so that you can

familiarise yourself. Is that OK with everyone?' They all nodded, amazed at the speed with which Tony had presented his idea. 'Right then, I'm off. Sally can get the printing organised now.'

John stood up as Tony left as quickly as he had come. 'Thanks for the details, Graham. I think it's shaping up nicely but I'll get back to you later if I may. Must rush now, I've got your consultants on my tail over at Abbey too.'

In the evening, Jan was keen to get on with sorting out the architect problem and they sat down with the brochure that had arrived from Chancery Architects. 'This is full of information but I've noticed that the one word that they've used throughout the brochure is 'planning'. It talks about agreeing the scope of the project before any money changes hands, vital stages in the planning process, understanding the implications of doing things in the right order. It even mentions contingency time.'

'Sounds good.' said Jim as he walked through the door and put his raincoat into the hall cupboard. He shouted to the boys 'Who does this pair of roller skates belong to?' Both boys came tumbling in from the garden.

'They're not mine!' Andrew shouted triumphantly. 'I'm keeping to the clean hall rule.'

'James. Move them.' ordered Jim.

'Come and sit down. I'll dish up the dinner in a few minutes.' Jan called from the living room. He subsided onto the sofa beside her. 'I think we should see this firm on Saturday. Look, they're open on Saturday at their office in the town centre.'

'Have you forgotten that I'm going to this Activity Day for work on Saturday?'

'Oh, hell, yes I had.'

'Can't cancel that I'm afraid. Tony Abbey's not impressed if someone misses a meeting.'

'But on a Saturday?'

'Afraid so. Can't be helped. Can't you get someone here any other time?'

'I'll try.' she muttered and stalked off into the kitchen to sort out the meal.

On his way into his office the next morning, Jim called in at the warehouse.

'Sorry to bother you, Alison.' He said, as he put his head around the door. 'I wanted to ask you about the special labelling that you do in here.'

She frowned. 'That causes a few headaches.'

'Why is that?'

'Well, once we've labelled them up for a specific customer, we can't sell them to anyone else, can we? If they get returned from a failed delivery, they're just stuck in the warehouse until we can persuade the customer to take them. There's lots of redundant stock on the shelves.'

'Ah, I see now. That's why Graham was asking me about Special Labelling.'

Alison put her hands together in mock prayer and laughed as she pleaded with Jim 'Please tell me that it's going to be sorted out!'.

'I think there's a very good chance that the change project will do that for you – or at least help to minimise the problem. It obviously causes you a headache – extra stock, taking up space – I know you warehouse people have a thing about space – but just imagine the problems that follow on from all that.' He checked them off on his fingers. 'Stock costs, space costs, we've made those goods when we could have been making something that is urgently needed, charges for failed deliveries, customer dissatisfaction... it goes on and on.'

'I know. Just think of the time that must be wasted rearranging those failed deliveries too. Customers must think we're stupid.'

'Exactly. That's a real cost.'

'So how will the new ways of working help with all that?'

'In lots of ways. It will highlight the actual amount of stock – and the costs involved. It will reduce the number of failed deliveries...'

'How can it do that?' she interrupted.

'By making sure the correct details are in the system when the order is first put into our system. Apparently, over eighty percent of our failed deliveries are due to dirty data.' Alison laughed but continued to listen with an intent look written on her face. 'We need to improve our data entry by an enormous amount.'

'Tell me about it!' she exclaimed. 'Some of the useless stuff – and missing information - on the computer drives me mad!'

'And,' Jim went on ' the special labelling is extra important because we can't get rid of it. It's a value added process - one that the customer is prepared to pay extra for, so it has to stay. We just need to make sure we do it right and that it isn't wasted.'

Alison looked pleased. 'I see the real importance now. It won't be easy to put it right though, will it?'

Jim paused in the doorway 'Never a truer word, Alison.'

Points to reflect on

- It is essential to influence the right behaviours
- Savings can be identified/benchmarked that will pay for new systems and processes. Develop a Return on Investment (ROI) model
- Identify the milestones and costs for the programme

Chapter 15 – Costs and Benefits

'From what I can see, Mrs Heswall, these plans are really very good.' Ian Hemingway, from Chancery Architects had arrived early for his meeting with the Heswalls and Jim was not yet home from the office. 'Perhaps you could explain a little about the problem while we're waiting for your husband.'

Jan had only just got in from work herself. 'Let's have a cup of coffee and we'll have a chat. And call me Jan, please.' She smiled to herself as she went through to the kitchen. 'Jim shouldn't be long now.' she said as she clattered the cups onto the tray.

'No worry. I'm comfortable here.' Ian replied from the dining room as he smoothed the drawings and examined the figures that went with them.

Jan placed the tray on the table and sat opposite the architect. 'Can I tempt you to a piece of cake? That should keep us going. As I explained when I rang your office, we've got a builder working on the project and we were hoping that he would co-ordinate the whole thing for us but we're not sure he's up to it. I'm beginning to get a bit uneasy about the planning. His attitude to planning just seems wrong to me.'

Ian looked directly at Jan. 'Certainly attitudes and opinions have a direct reflection on performance so you're right to be worried. Let's start at the beginning. What are you trying to achieve with the new house?'

'Ooh, that's a big question. Let me see. Obviously, we want a new house. One that will be big enough for all of us. Jim and I are recently married and we've four children between us. Jim says it's like a company merger so we need reorganising.'

'That's a good start. That's your impetus for the move. Obviously, the impetus needs to be compelling because you'll be spending a lot of money and it won't be easy. It's a big project, so something must be driving you. Was it not possible for you to stay here?'

'We did consider extending this place – a conservatory, and so on – but we really needed another bedroom and at least one more bathroom plus an ensuite so we decided to take the plunge and get our own house built.'

'Fine.' He nodded towards the framed photograph on the sideboard. 'I guess they're getting to the age where they want their own space, aren't they?' he said.

'Yes, that's the problem. Trying to get into the bathroom in the mornings was driving Jim crazy.' She stood up and went to the window. 'Speak of the devil, here is now.'

The front door opened and Jim shouted 'Sorry I'm late.' He rushed into the dining room and shook Ian's hand. At Jan's questioning look he said 'No, I won't bother with coffee, let's get down to business.'

'I've been explaining to Ian about the bathroom problem.' She turned to Ian and asked 'Is that an impelling impetus for change?'

'Seems like it to me.' He smiled. 'So, what's made you want to change what your current architect is doing? These plans were obviously produced by someone competent.'

Jim stepped in. 'It's not his design or building work that we're worried about. It's just the planning of the whole thing. We get the distinct impression that he isn't planning ahead. We're worried that this will lead to delays and increased costs and possibly other problems that we couldn't even begin to imagine.'

Jan took over. 'We just want to feel comfortable with the whole process. And we don't.'

'Well, I can't promise a totally smooth ride but I can see now what you're getting at. It might help if we do a cost and benefit analysis here.'

Jim looked at Ian in surprise. 'That's exactly the stage we're at in our merger at work.' He looked triumphantly at Jan. 'I told you remarriage was like a company merger.'

'Yes, your wife has already mentioned that. I can see what you mean.' He took a pristine piece of white paper from his black leather folder and neatly drew two columns. 'Costs... and benefits... Let's see, more family space, extra bedroom. Those are the solid, obvious things you will get out of your new home but can we put something more that you'll get out of a properly functioning family home?'

'A happy family maybe?' suggested Jan with only a little hesitation, while Jim sat unmoving, looking bemused.

'That's the sort of thing.' encouraged Ian.

'Getting to work on time every day is what I'm hoping to get out of it.' pronounced Jim as he started to understand their reasoning. 'I think I'm just catching up with you two now, Ian.'

'No problem. And am I right in thinking that if you get to work on time you'll help your job security?'

'That's right.' said Jan 'so that's a definite benefit for our family. It also gives our marriage a better chance, doesn't it?' She smiled at Jim while Ian continued to write. 'Let's just go through some of these costs now, then I can enter them into my analysis.' They worked on until they had arrived at a point where Ian had all he needed and Jan and Jim had had their reasoning reinforced. The final result would be worth the trouble and expense. 'Right, give me a few days and I'll send you a quotation and my thoughts on the project. Then we can see where we're going.'

Jan was smiling as she closed the door and Jim knew that she was happier with the project than she had been for a few weeks now. 'That was good. Let's collect the kids and go out to eat. My treat.' he suggested. She smiled even more broadly.

The following morning Jim had no problem getting into the bathroom. The children were all having a lie in. No school or activities this Saturday morning. 'I should be home late afternoon. I'm sorry to spoil your weekend but this event is really important. I was glad

that you were able to get the architect here last night. I'm looking forward to seeing his proposal.'
'Me too. Should be interesting. Don't worry about our day. I've plenty to do here and the kids are going to tidy their rooms.' she pronounced.
'My, you looked a bit fierce when you said that. They won't argue with you!' He kissed the top of her head and left her still at the breakfast table.
He quickly found John and Caroline in the foyer of the hotel that Tony had hired to host the first part of the Motivation Day. He almost didn't recognise either of them in their casual clothes. 'Good morning, Jim. Caroline and I were just discussing the purpose of this thing.'
Jim smiled. 'Good morning, John, Caroline. I'm not sure what to expect but I guess it's an attempt to get us all behind the change project. Tony obviously wasn't impressed with the turnout at the last Steering Group meeting.'
'You're right, of course. He needs to broadcast his message a bit too, being the Project Sponsor.'
Caroline was listening with interest but, at the same time, looking around her. 'It's a good turnout anyway.'
John laughed, but it was a laugh devoid of humour. At that point, Graham joined them, preventing John from making further comment. 'Morning all.'
Morning Graham.' said Caroline 'I was just saying, it's a good turnout.'
'Yes, you're right – but we wouldn't expect anything else, would we? Haven't seen Ron from your warehouse though.'
'Oh, no. That won't go down well.' She looked concerned.
Grand double doors were opened at the far end of the brightly lit foyer and Sally Osborne, Tony's assistant stepped out of the large meeting room, which was set up with circular tables for eight people each. 'Good morning everyone. Would you come through please? There's coffee in here for you.'

Several people did not hear her low-voiced announcement but merely followed the crowd as they all shuffled into the room. They hovered on the threshold, not knowing whether to sit or stand. Tony bowled into the room and, in a loud but cheerful tone, welcomed everyone to the event.

'Get your coffees and sit down, please folks. Just choose a seat. Let's get this show on the road.'

He waited with carefully disguised impatience until everyone was seated. This took several minutes. 'Right. I'm glad to see you all and grateful that you have given up your Saturday. I'll try to make it easy-going. The purpose of this event is to help us all to work as teams and for us all to have a bit of fun at the same time. I want to encourage you all and to make sure that our Business Change Project goes as smoothly as possible. I know it's been said before but it's worth repeating that this project is absolutely imperative. Without a successful project, the merger of our two companies will fail. If that fails, then we all fail.' He raised his voice and the passion could be felt almost physically in the large, vaulted room. 'I am not prepared for that to happen. Make no mistake, the downward spiral of our companies fortunes will continue until we have examined the business systems and processes at Abbey and MML and taken steps to put the situation right. Any questions?' There was silence in the hall. 'I'm sure there must be one or two questions. Anyone got any concerns?' He sat down and looked around expectantly.

John stood up. 'Can you outline what might be necessary?'

Tony smiled. It was as though John had fed him just the right question. 'Thank you John. I'm sure a lot of people must be wondering that. The short and simple answer is 'Whatever it takes' but broadly speaking we need to get through this ERP implementation, making as many of the right changes to the processes as possible along the way to increase our efficiency.

Increases in efficiency will lead directly to increased chances of survival. First though, I know that many of you might not have met each other before so I'll hand over to Graham – our chief consultant on this project. I think he's got a few ways to get us all up loosened up.' Tony sat down and Graham swung into his well-practiced introduction routine including a couple of icebreakers for the meeting so that everyone felt more comfortable. He knew that it was vital that people felt free to give their comments and ideas.

As Graham had agreed with Tony, Jim and John before the meeting, Graham led a facilitated 'brainstorm' to bring out a few questions that everyone needed to appreciate the answers to. His voice, and the challenging questions he was asking, made everyone sit up and listen while his humorous approach made them forget, at least for a while, that they were working on a Saturday morning. 'Right, let's split into four groups. Here are your group facilitators.' He pointed to John, Jim and Tony. 'That's three of them and I'm the fourth one. Don't' spare our feelings. Just make sure that you have a good discussion of the questions I've given you.'

Jim's group included several of the sales staff who were keen to put their point across when Jim asked 'What makes a successful company? Give me some examples of best practice.'

The Sales Director from Abbey, who Jim had met several times and knew was well respected, spoke up 'Imperial Engineering have a good reputation and I know that they have a professional approach. Their sales staff have all the information at their fingertips. They can give a customer the answers he needs on the spot rather than having to go back to the factory and search it out like we have to do. Will this project help us that?'

Jim beamed at him. That was just how he wanted this to go. 'It certainly will. Information is key to this whole thing. Let's just look at that in a bit more detail then we'll go on to the other questions.'

In John's group things were not going quite so smoothly. The customer service staff were not quite as forthcoming but eventually John was able to bring out points about how the new customer service working would include more authority to deal with queries and complaints. 'We're talking about empowerment here. Does that sound good to you?' This was met with an excited buzz of conversation and agreement so he moved swiftly on to question them on the advantages of a company culture where there was no fear of censure and where their suggestions and problems would be listened to.'

All four facilitators were beginning to enjoy the sessions but Graham looked at his watch and knew he had to move them on so that they covered his planned programme. He called them all to order and continued 'Now, we've a few more ideas up our sleeves to help things go smoothly. One that you will see starting with this morning's meeting will ensure that our attitudes are working with the project rather than against it. All our meetings from now on will be subject to the use of Red cards and Green Cards.' He signalled to Sally, who walked quietly to the back of the room where a large box was stored. 'These will encourage the right changes to our behaviour in meetings. We can't continue to have poor attitudes towards the changes. We simply can't afford to fail.'

By this time Sally had placed a small pile of the laminated cards on each table and they had been passed around. Muttering had started.

Graham raised his voice. 'Let's just look at the cards. We can all have a chat about them later and I'll look forward to your feedback.' He picked up a card and held it up so that the red side faced the sea of faces raised expectantly towards him. 'Look at point number three, ladies and gentlemen.' he ordered. All the heads dropped as they read 'Don't interrupt someone when they are making a point.' Understanding dawned on several faces and the silence held. 'Okay. Let's just look at two important points on this card. Stay on the Red side for a moment. Point number one – 'You will not criticise the company or the project'. Remember that. There was total silence. He turned the card around and solemnly showed the green side to his audience. 'Please read aloud with me the first and most important part of this card 'We can do it!' The few people who joined in were almost inaudible. One more time, please 'We can do it!' This time there was a marked increase in the enthusiasm and the volume and as Graham sat down, Jim stood up and started to applaud. Soon everyone had joined in. Some people looked amused, some elated and some bemused but they all joined in the applause.

When the noise had died down, Jim spoke 'Well, that sorted a few things out. We now all know why we had to make this change and I hope you'll all join in like that at the workshops we'll be running in the weeks and months to come.'

Graham stood up again. 'We've got a little film show for you now, folks. As you might expect, it's not purely for entertainment, but it is certainly amusing. I think you'll enjoy this little lesson.' The lights went down and for the next thirty minutes or so they were treated to a video of 'An Imperfect Implementation.' The scenes of disaster which followed the characters' behaviour were exaggerated and, at times, farcical, but the lesson hit home. When the lights went up again, an animated buzz of conversation went up.

Tony banged on the table to get a reduction in the noise levels so that he could be heard announcing 'Lunch is served.'

Lunch was a long and relaxed affair with everyone sitting at the circular tables with a mix of people from the two companies and from different departments. All the senior personnel, including Jim, John and

Graham had been briefed to 'spread themselves around'. Jim found himself sitting at a table with several members of staff from Abbey who he had not met before while John made sure to sit with people from MML's HR Department. One young woman with a cheeky grin and piercing blue eyes commented 'This is cool. I thought there would have been a seating plan and we would have to sit with a load of old men.' She looked around the table at her fellow diners and stopped as she caught John's eye. Her cheeks went bright red and she continued hurriedly 'You know, forced to socialise with boring people.' She looked down and concentrated on her meal.

John grinned 'Yes, these events can be a bit of a trial, can't they? Always having to watch what we say!'

Sally had put together random team lists for the next stage of the day. She gave a Team List to each of the senior staff with instructions that they were to gather their team as they strolled down the road to a Ten Pin Bowling alley, leaving their vehicles behind on the hotel car park. This proved to be a challenge to all of them as they had not met some of their team members. It met its objective however. By the time they had piled into the reception area of the Bowling Alley and organised their footwear from the rental counter, they were all chatting and laughing as though they had known one another for years. Rivalry between the teams was fierce as they made their way to the scene of the afternoon's action.

The following Monday, Jim was pleasantly surprised to find that the atmosphere was noticeably improved. The teams had become closer and seemed to have taken on board the lesson that working together produces results. He also reflected on Tony's active sponsorship of the project. If the CEO had not dealt with the problems of motivation promptly then the programme would be damaged. Its chances of success would certainly be reduced. He was impressed with how Tony had not let the problem drift on but had put time, effort and a quantity of cash into it to put it right. To drag the project back on track. He was keen to get started on his people-focused workshops using the new system of Red Cards and Green Cards – it was just the tool he needed to get the message across about improved attitudes. However, this would have to wait until Friday. Before that there was Graham's Costs and Benefits Presentation to Tony and Neil. It would be really interesting to see the final figures and to witness the reaction of the two CEOs.

The following day, the four men were seated around the boardroom table with Graham at the front of the room with his laptop connected to illustrate his presentation. Jim looked at the Senior Consultant and reflected that this was a big moment for him. If Tony and Neil were not sufficiently impressed and convinced of the outcome then the change project could stop right away. Jim found himself offering a silent prayer that Graham's ideas would prevail and it wasn't until that moment that he realised how much the project had come to mean to him.

'Good afternoon everyone. This is where we learn the figures behind the obvious benefits – and the costs – of our Business Change Project.' started Graham. 'You are all, I'm sure, painfully aware that failure is not an option. The main benefit that we are aiming at is one merged, and recovered, company.' He pressed the button on his laptop control and the screen flickered into life. A graph showing the declining fortunes of both Abbey and MML appeared. These figures were obviously not a shock for either Tony or Neil but they still made scary reading. This was quickly followed by the new company name and logo. 'When we agreed this change we were aiming at looking and sounding professional, weren't we?' he asked. There were murmurs and nods around the room. 'OK, how are we going to make the profit figures match that image?' He brought up details on the screen showing the ways he planned to help. There will be plenty of rationalisation – of products, of customers and suppliers. We will automate processes where possible, get rid of duplications, improve efficiencies, get better utilisation of employee's time and less absenteeism due to stress. We will improve quality, increase productivity and reduce inventory. And, of course, implement the software to support all this.'

'That's an impressive list, Graham.' said Tony. 'Do you have figures to go with those?'

'I certainly do.' answered Graham and proceeded to the next stage in his presentation. He showed them tables full of figures demonstrating enormous possible savings that would result from implementing each of the modules.

'Impressive again,' commented Tony 'Can we see a bit of detail as to how you've worked those figures out. The effect of implementing the Purchasing and Planning Modules, for example?'

'Of course, we can go through them all, then I'll give you printouts and you can study them at your leisure.' He ran quickly through the figures in the Planning Module, showing all four men how inventory would be reduced and that this would result in reduced working capital. He also turned to the Purchasing Module and showed how they could start to control suppliers and save money on almost everything that the company purchased.

'You'll no doubt have noticed that the Manufacturing Module has been left out of the costs. Do you want to say a word or two about that, Tony?' asked Graham.

'Yes. Thanks, Graham.' Tony looked around as all eyes turned towards him. 'As you all know, I've been reviewing the budget as we go along and I've come to the conclusion that we have to do a certain amount of descoping.' He saw the concerned looks on their faces and hurried on. 'Just temporarily, of course. We have to put the Manufacturing Module on hold for now. It's an enormous module and Graham and I have come to the conclusion that the time that would be necessary to do the job properly is just not available right now. I know we've carried out a lot of work on that module but most of it has been helpful in sorting out some day-to-day problems and also in providing information for the other modules. The conclusion is though, that continuing with the Manufacturing Module would be to the detriment of the other modules. The cost of a huge module like that is also an issue at this stage. We just can't afford it.' He glanced around at their subdued faces. 'Nothing to worry about. It will happen eventually. Let's get other parts of the business in order first.' He indicated to Graham to continue.

Graham resumed control of the meeting. 'The positive side to this is that we can concentrate on the other modules, get them right and then use the benefits from them to improve the implementation of the Manufacturing Module when the time comes. Hopefully, it won't be too long before that happens.'

Neil spoke up for the first time. 'Sounds good to me but how much will all the rest of this cost? We've paid out a fortune already.'

'Well, it's true that we're well into the project in some ways. We've done a lot of data gathering and detailed design already and this has produced some early cost savings but there is still a lot of hard work to do before we get into the real cost reductions. Let me show you a cost summary.'

As the screen showed the figures, including the cost of thousands of hours of training, the hefty software costs and the consultancy fees, there was a sharp intake of breath. Tony, Jim and John looked down at the table as Neil got to his feet, his eyes widening. His eyes seem to be on stalks, thought Jim, and I'm not sure that this is good for his blood pressure. Neil slumped back into his chair, breathing deeply, and looked at Tony. 'This is just what you expected, isn't it?'

Tony looked at the older man with sympathy. 'It is indeed. We knew something drastic had to be done – and this is it in my opinion. Let's just get to the end of Graham's presentation, then we can discuss this further.' He indicated for Graham to continue.

'Thanks, Tony.' Graham carried on, gaining in confidence as he got nearer the end of his figures. 'I honestly can't see how else you could achieve the savings and the drastic changes in working practices that are absolutely vital.' He sat down. 'Any questions?'

John looked towards Tony and Neil. 'Maybe now would be a good time to take a break?'

'Yes, I think so. Let's reconvene in thirty minutes. Is that OK with you Graham?'

'That's fine.' Graham gathered his files, handed out the prepared folders to the four men and left the room.

Jim and John kept quiet as Tony and Neil had a heated discussion. Tony maintained his usual optimistic stance in the face of Neil's resistance to spending the huge amount of money. John, however, could see that his boss's patience was running out. Eventually, Tony declared quietly but with obvious determination and seriousness 'There is no other way, as far as I can see, Neil. If any one of you can come up with a viable alternative, I'm listening.' He looked first at John, then at Jim (for what seemed like a lifetime to Jim) and finally at Neil. His stare was calm but cold. Nobody said a word. Jim found the silence extremely uncomfortable but dare not move. 'Right then. We go with this. Increased profits, here we come.'

Graham came back into the room a couple of minutes later. Jim observed that he seemed to have recovered his composure and appeared to be confident of the outcome. 'Do we have a decision?' he asked.

'Yes, we have a decision. I'm sure you know that we must go ahead. Let's do it.'

Points to reflect on

- Comprehensive data /facts, are essential for good analysis and decision-making
- It is not possible to do everything at once. Prioritise the project
- Focus teams on staying positive
- Simplify expectations and communicate expected behaviours of the teams
- Quantify how the benefits add value to the organisation going forward

Chapter 16 – Identify main impacts of change

Friday morning saw Jim, Alison and Meg sitting in Jim's office making a start on producing the next edition of the newsletter.

'Thanks for coming over, Meg. Perhaps we could hold the next meeting over at your place. It would give Alison a chance to get to know the rest of this new company a bit better.' suggested Jim. Alison's eyes lit up and Jim reflected, not for the first time, what a difficult and frustrating time she must have had working on a boring job in the warehouse where initiative and enthusiasm had not been valued or encouraged. How much more talent and enthusiasm was there, just waiting to be discovered? Poor systems with a lack of information available obviously produced mind-numbing jobs. Not good for anybody. He dragged his thoughts back to the job in hand. 'For now, let's get this show on the road. What we're aiming at is some means of communicating with all the staff. The official brief is 'Ensure staff understand the compelling imperative for change and encourage creative input from all areas.' We obviously need to understand what steps we will take and what their role is. I suppose the message is 'change is coming – get ready!' Any thoughts?'

'Yes, lots.' enthused Alison. 'We must make sure that everyone feels part of the organisation and that they see the benefits that better systems can produce.'

'Mmm...Motivation.' mused Meg.

'Yes, that's certainly what we're aiming at but how?' Jim asked. 'One thing's for sure, we can't do it all ourselves. We need to make sure that other people contribute.'

'Perhaps we could have a guest writer for each month's newsletter? Maybe start with Graham giving a few pointers about what he's trying to achieve? Do you think he would he do that?' suggested Meg.

"Let's not give him a choice!' laughed Jim. 'I'll tell him what we need. I think this edition should also include a piece from Tony as the Project Sponsor.'

Alison had been doodling on her pad. 'I think the newsletter title is a bit boring. Abbey Martin News?' Her face showed her incredulity. 'Couldn't we call it something else?' she asked. 'What about 'Phoenix'? You know, rising from the ashes.' She looked at Jim and Meg in turn and saw that her suggestion had met a receptive audience.

'That's great.' Jim declared. 'Let's go with that. Now, we just need to share out the practical bits and pieces between us and we've got the makings of a really good motivational tool for the Project.'

On Saturday morning, yet again, Jim, Jan and the family were on the building site. They had arranged to meet Barry there and they all knew that it wasn't going to be a comfortable meeting for several reasons. The major reason was Jim and Jan's growing concern and dissatisfaction with Barry's performance. As they squelched across the site, another problem became increasingly obvious. It was wet. Very wet. There were no workmen around and the trenches that had been dug in the previous week were full to overflowing with thick, black water. The rain still poured down and Jan hung on to Jim's arm to keep her balance as her feet stuck in the mud, as well as keeping close so that she could share the large umbrella. The boys leapt from puddle to puddle, while the girls delicately avoided any chance of getting dirty, lingering on the edge of the plot. The adults picked their way amongst the foundations and markings that showed them where their new home would eventually be. 'Isn't this where the conservatory will be?' asked Jan, looking at a large oblong-shaped stretch of concrete sunk into the mud. James began to race around the outside of this concrete and Jan looked at him thoughtfully. 'We haven't included any landscaping of the garden area in our plans, have we?' she asked.

'Apart from just turfing the whole area, no, we haven't. We agreed that we couldn't afford to do that until everything else is finished – and paid for.'

'When we get round to the garden, we're having a pool. Aren't we, Dad?' declared Andrew. Jim silenced him with a look.

Jan answered him. 'Well, perhaps, Andrew but what I was thinking of was a path around the conservatory. It seems a bit silly to lay grass there and then dig it up again. We might as well put a proper path in place while the garden is already in a mess.'

Jim nodded. 'You're right. That will probably save money and trouble in the long run. Let's mention it to Barry.' He looked at his watch. 'When he eventually arrives, that is.'

'Look at the mess on my boots, Mum.' moaned Emma, picking her way through the mud to her mother's side. 'My hair is getting wet too. Look it's going frizzy. Is there anywhere to shelter?'

Jan sighed. 'Emma, it's obvious that there's nowhere to shelter. This is a building site!'

'Jim handed her the keys to their car. 'Here. Go and sit in the car for a while.'

'Don't get the inside of the car muddy.' Jan shouted after her daughter. 'Oh, here's Barry now. About time too!'

'Sorry I'm late. Got held up on another job.' He said as he rushed over to Jan and Jim. 'I thought we might have got a bit further along with the house by now but you know how things go.'

Jan's face darkened. 'No, I don't know how things go. It seems that things don't 'go' at all! Why don't you tell us what's been happening?' She glared at the builder.

'Well, nothing much. It's the rain you see.'

She took a deep breath. 'Yes, we can see it's raining, we've been waiting here, getting wet.'

Barry looked nervously at his watch. 'Yes, sorry about that.'

Jim hurried into the conversation. 'You're here now. Let's hurry up and review the progress. Has there been any?'

Barry hesitated. 'It's the weather. Can't do anything about that. It will mean completion will be delayed though.' He fished in his pocket and pulled out a crumpled piece of paper torn from an exercise book. He smoothed it out with difficulty and tried to keep it dry as the rain continued to fall. 'Yes, it means we'll have to delay this stage...' he pointed at a wet and blurred item on the paper. He retrieved another grubby paper and held it out to Jan. 'This is your list of milestones that you asked for.' He blinked repeatedly as Jan and Jim looked first at him, then at each other.

'I think we've got a problem here...' began Jim.

Barry stuffed the soggy paper back into his pocket. 'I can get my wife to type the list out for you if you like.'

Jan could contain herself no longer. 'That won't solve the problem. The problem is the lack of progress and the lack of proper answers from you, Mr Bryson.'

He stuttered. 'But, you didn't tell me...'

'Tell you! Tell you what? That it might rain, that we wanted this house built on time, or that we expect a professional attitude?'

Jim took his wife's arm. 'Come on, there's no point in getting upset – or in us getting any wetter. Hurry up boys, get in the car.' He turned to the builder. 'We'll be in touch.'

The family spent an uneasy weekend deliberating the possibility of sacking Barry Bryson altogether rather than, as they had first planned, keeping him on and perhaps engaging Ian at Chancery Architects to oversee the project. They waited impatiently for Ian's quotation and it arrived just as Jim was leaving for work on Monday morning. 'Can't hang around, Jan. I've got a few attitude problems with people at work as well as on our house building project.' he yelled as he raced out of the door. 'You have a good look at that and then we'll discuss it tonight.'

'Don't be late home then.' she said, as she kissed him goodbye.

'I won't, I promise.'

As he drove along he tried to remember what Ian had said about attitude. Whatever it was, he knew it had applied to Ron Whitehouse in the warehouse. Neil had asked him to sort out the situation with Ron. Jim knew that he didn't really have any choice. Ron and the Project Committee had to part company. He couldn't take up a valuable space on the Change Committee. Ah! That was it! 'Attitudes and opinions have a direct reflection on performance' Well, there you are, he thought, that's Ron's problem.

'So that's it, I'm afraid Ron.' He looked across at the older man and had to admit to himself to feeling some sympathy. Change was never easy but it must be especially difficult for this man who had spent the bulk of his working life without the aid of computers and indeed, had done his best to avoid them over the last few years. 'We really can't afford any passengers on the Change Committee. There's too much work to do.' He was not surprised to see relief wash over Ron's face.

'I'll not pretend I'm sorry, Jim. You know me better than that, it seems, but who could possibly do the job instead? You need a representative from logistics here at MML, don't you?'

'We certainly do. And don't forget that the company is called Abbey Martin now.' He watched with impatience as Ron's face took on its old, stubborn look. 'I rather think Alison would fit the bill.' Jim was pleased to see the surprise on Ron's face as he stood up and opened the door to show Ron out of his office. 'Don't mention this to her, Ron. Just ask her to spare me a few minutes as soon as she can.'

As Jim stood by the door, Ron sat immobile in his chair, not speaking. Jim returned to his desk and Ron started his tirade. 'What do you mean 'don't mention this to her. She works for me. Did you hear me, she works for me so I will be the one to decide if – and I repeat if – she will take on any extra work. I must say I'm surprised that you think she can offer anything to this committee of yours but she works for me so I'll let her know what she has to do and I will decide when too.' As he spoke his voice had gradually raised until he was shouting at Jim and as he finished his outburst he sat back heavily and crossed his arms over his chest, the usual stubborn look on his face.

'Have you finished?' asked Jim quietly. Without waiting for an answer he continued 'This is a company project, Ron, and I'm the Project Leader. It is totally irrelevant which department Alison works for. If I say she will replace you on the committee, then she will and the way I deal with it is up to me. I will not be asking you for your approval or having you control the timing of it. Now, if you don't feel you can cope with that, we can speak with Tony right now.' He looked questioningly at Ron who stared at the floor, breathing heavily. Jim then stood up and, with as much patience and control as he could muster, strode over to the door and held it open for Ron who pushed past him without another word.

It was a heavy week for Jim - people-problems occupied his every thought at work. First Ron, now his first motivation workshop using the new cards. He had chosen to try it first with his erstwhile colleagues in the Purchasing Department. He knew the subject matter, of course, and he was interested to see how they had tackled the issues involved in redesigning the processes. He also thought that it would be fascinating to get to know their counterparts who worked in the old Abbey Purchasing Department.

They filed into the meeting room, their faces wary. They did not know what to expect, thought Jim. He was determined to start with the green card – this was a workshop about motivation, after all. The initial introductions over, Jim asked a few questions about how the project was progressing and was relieved that several people were confident that the change would be successfully accomplished. He immediately showed them the green card and announced the reason 'You adopted a positive attitude.' He immediately swung around and showed the green card again, this time declaring 'You were actively listening'. This is easy, he thought to himself. Then he realised that he couldn't remember many of the statements shown on the red card. He read it again and had just got down to the last point when someone's mobile phone rang. Aha! The red card shot up. 'We shouldn't answer phones in meetings unless we have cleared it with the other participants first'.

All faces turned expectantly towards Jim. 'Right, I had better get on with this meeting or I will have to show myself the red card for running a meeting without clear objectives.' He handed round the preparation documents for the new Purchasing System – the result of the preliminary data gathering done by Graham and his fellow consultants. 'Now, what we want to achieve is continuous improvement so let's not accept this as a finished process. We're all Purchasing experts here so let's look at this with a practised eye.' The papers rustled and a self-satisfied murmur made its way around the table. 'First item, we need to take the inefficiencies out of purchasing orders. We need to take out as many manual steps as possible – the things we don't need to do.' A sea of blank faces met this remark. 'I'll give you an example.' said Jim 'Look at this on page five.' The heads went down. 'And page six and seven. It happens throughout the process - people keep signing things. That is just not necessary. We can set up pre-authorised levels on purchase orders. Let's get going.' They set to work and by the time the meeting was over, they had streamlined the purchase order system to an even greater extent than the consultants had managed.

They had also discussed and set authority levels and developed a system to note accruals to tie in with the accounting periods. They even had elements for a next phase – e-procurement.

As they left the meeting room, Jim handed them all their own copy of the Red and Green Cards, telling them 'Feel free to use them whenever you're working together. And let me know how useful you find them.' The laughter could be heard receding down the corridor as they joked about the various items on the two lists and flashed the cards at one another.

Jim was relieved. Another hurdle crossed. He walked over to Graham's office to discuss the progress and was surprised to see Tony already ensconced there.

'Come in, Jim.' Graham answered in response to Jim's tentative enquiry.

'Hi, how are you doing?' asked Tony.

'Going great guns.' replied Jim.

'Great. Any news?'

'I've just finished my first workshop using your card system. It went really well.'

'Well, that is good news. When is the next one?' smiled Tony.

'Hey, slow down a bit.' he laughed 'I was going to suggest we ran through some of these Purchasing processes, Graham,' he said, indicating, under his arm, the document files that had just been completed in the meeting. 'but if you're busy, I can see you later.'

'No.' said Tony 'I was just going. I've got to see Neil before I go. See you later.' He left with his customary hurry.

'I'm glad it went well. Something has to go well around here.' proclaimed Graham.

Jim immediately flashed a red card and summoned a smile to the consultant's face. 'No negative throwaway lines!'

Graham held up his hands. 'Guilty as charged. It went well then? It worked?'

'It certainly seemed to. We got through a heap of stuff on the Purchasing processes – put in a lot of automated workflow. It's looking good, particularly in the area of authorisations.'

They went through the meeting documents in detail and Graham pronounced himself pleased with the progress being made 'I'm also really impressed with the way you're coping with the personnel side of things, Jim,' he commented 'and we can definitely see that a lot of people here have taken on board the principle of continuous improvement.'

Jim smiled. 'I think Tony made it perfectly clear that we don't have any other option than to make this thing work. Most people – with just one or two notable exceptions – are putting a lot of effort into this.'

'You're right. I think there's a lot going on beneath the surface though. I don't think all is going smoothly between Tony and Neil. Bit of a power struggle there, I think.'

'What makes you say that?'

'Oh, nothing that Tony said really,' admitted Graham 'just reading between the lines.'

Jim could see that he was not going to get any more out of the consultant on that topic at this stage so he changed the subject. 'Sounds intriguing, you must keep me in the picture. What else is going on with the Project?'

'We've got a bit of scope creep coming up, I think.'

'Scope creep. What does that mean?' laughed Jim 'Wait, don't tell me. It's just what it sounds like, right? The scope of the project is changing. What are we adding then?'

Graham consulted his notes. 'It's 'available to promise' or 'ATP' as we know it. An add-on to the Order Management module. It's where manufacture of stock is planned so it's available to promise for eventual delivery to customers but it isn't actually physically in stock yet. Should be useful.'

'This Project changes all the time, doesn't it. It never stays the same for two days together. First we had 'descoping' now we've got 'scope creep'. It's definitely just like building a house. We've got lots of changes there too. We've added things and subtracted them – I just hope we don't have any personnel changes!'

Graham laughed along with Jim. 'I thought that was what your Home Change Project was designed to avoid!'

Jim was careful to get home on time that evening. He knew he had to attempt to raise the motivation at home as well as at work and also that Jan would be impatient to discuss the proposal from Chancery Architects.

She was so impatient that she met him at the door with the letter in her hands. 'It's not really much more expensive than our friend Barry's.' she said, waving the quote under his nose.

Jim walked past her into the hall where there was a surprising lack of boots, shoes, toys, coats and bags. He didn't have time to reflect on the efficacy of the clean hall policy as Jan was hot on his heels. 'Give me a second, Jan. It's been a hard day. Anyway, how expensive is not much more expensive?' he asked.

They sat down at the dining table and Jan rushed on 'The consultancy fee is extra, of course but it will be worth paying that just for the peace of mind, won't it? And some of the labour costs are up a little but the materials costs are very similar and even lower in some areas. There are a lot of clauses about maximum variations, there's even a list of review points – I suppose that would be equivalent to a framework for our Milestone Planning.'

'Phew. This is a lot to take in but it looks good. Let's have something to eat and then we can settle down and review this properly.'

Jan looked blank. 'Oh, I haven't had time to get a meal organised. Can't we send out for something?'

'Do we have a choice?' Jim remarked, going through to the kitchen to put some coffee on. When he went back to join Jan, she was on the phone already, ordering food.

As soon as she came off the phone, she was back on to her favourite subject. The house. 'Ian says...'

'It's Ian now is it? He seems to have got his feet under the table quickly enough.' Jim joked. He saw Jan's look of impatience and tried harder to concentrate 'What does Ian say?'

'Look, it's here in his letter. 'Dependent on hiring our own, trusted sub-contractors to carry out all the work carried out in Appendix B.'

'Let me see. What's in Appendix B?'

'It's all the main construction work plus plumbing, electrics, gas supply and so on. It also includes the kitchen installation. Now, that is something I need to speak to them about. Barry was going to let us hire our own kitchen people. I must check that we can choose exactly the layout and style we want.'

Jim looked thoughtful and concerned. 'It won't all be perfect, you know. You might have to compromise a bit.'

They spent most of the evening going over the Architect's proposals and Jan made a list of the questions and issues she needed to raise with Ian Hemingway.

'You know, I have a good feeling about this now. I feel easier than I have for ages. A few more months and we'll be living in our new house. It will have all the space we need. Just imagine!' said Jan with a dreamy look on her face.

'Well, if you're happy then that will be a major advantage.' he commented. 'We might even get some home-cooked meals in this wonderful kitchen that you're planning.'

Points to reflect on

- Be aware that Scope Creep may happen and must be controlled
- Look for continuous improvement and quick wins
- Detailed – and repeated - evaluation of the workflow processes will produce results
- If there is evidence of lack of commitment, take people off the project – they cannot be allowed to 'fester'
- Ensure that the people who have potential to be influential are fully on-board and manage the change impact
- Start early on your transition plan from 'as is' to 'to be', a measured, phased-in approach or a go-live cutover

Chapter 17 – Agree scale and phasing of changes

'These meetings come around with alarming regularity.' sighed Graham as he sat down at yet another Steering Group meeting.

Jim was chairing this month's meeting and he moved swiftly to ensure that it did not become a session dominated by moans about workloads or problems.

'Let's welcome Caroline. Making a guest appearance for the HR Team for the first time at one of these meetings.' He smiled at her and she nodded back, obviously uncomfortable.

'Right, the main business today is to make recommendations for the scale and phasing of the changes. We need to decide which module will be implemented first and – just as importantly – when. Graham, can we have your thoughts on this please?' He looked across at the consultant who walked without delay across to the laptop set up to show details of his presentation.

'As you are all aware, we're still heavily into the Design Phase and there's still an enormous amount of work to be done in most departments. However, that doesn't mean that we can't start implementation of some modules in the very near future.'

He ran through his presentation. 'I would like to leave the Manufacturing Module to one side for now. That's under special consideration isn't it, Tony?'

Tony merely nodded. Unusually subdued.

'What I am suggesting is that we start with the Planning, Purchasing and HR Modules. HR is usually quick and easy to implement but might not show noticeable benefits, Purchasing is a fairly easy module and Planning would give us real benefits. There is a limited amount of configuration needed with these and this will help Frank and his colleagues who are dealing with the detailed technical design. No doubt when they've got these modules successfully configured, they will be up to speed on all sorts of things here.' He smiled across the large table at Frank Littleton. 'Of course, this will be a sort of trial run for all of us – not just the software people - and there's a definite advantage for us to start with these modules. A big implementation can bring big problems, so let's start small.' Everyone nodded in agreement and Graham continued. 'A couple of other things have informed this decision. Firstly, that the Order Management and Finance Modules absolutely must be implemented together. We can't run one without the other so we don't have a choice there. Secondly, it means that we can start some implementation relatively quickly and get some benefits from this project as soon as possible. Are you OK with that, Caroline, on the HR side?'

''I'm obviously a bit apprehensive about it but I'm also looking forward to seeing the advantages of a fully integrated and automated system. We've struggled with separate systems in HR for too long. It can take ages to resolve queries. It just doesn't work at the moment. I have no confidence in the information that I have to supply for all sorts of reasons.' Her voice dropped to a conspiratorial tone. 'Not least of these is an incident we discovered recently where we'd been paying someone for two years after they had left the company. Really!' she exclaimed as she saw the looks of disbelief on her colleagues' faces. 'Everyone in the department gets very stressed when we know things are not as they should be. So, yes, let's go with the new system.' 'Thank you Caroline.' responded Graham 'I think we get the picture.' Everyone appeared to be in sympathy with Caroline. They had all been in a similar position.

Tony broke his silence. 'I think Caroline has just made a very valid point. We've concentrated on how the business needs accurate information to make effective decisions but we shouldn't underestimate the impact that being unable to do our jobs properly has on all staff. Caroline has highlighted this. Poor information not only leads to problems in the business but it affects us all.' Yet again, several heads nodded in agreement.

'Right then,' Graham continued 'to sum up, we'll go with the HR, Planning and Purchasing Modules. From the end of this month we will start configuration, then testing, then data conversion into the new format. The exact date for going live will be notified as soon as Frank and the rest of the Steering Group give me the go ahead.' He looked across at the software expert 'Frank and I will get the detailed planning for this part of the implementation finalised and we'll all carry on with the data gathering and design. As you can see, the different modules will go forward at different paces. This whole phase is long, sometimes boring and always messy and, I've got to warn you, it will last for several months yet. But, it will be worth it, I can promise you that.' He sat down and met the eyes on plenty of smiling, but slightly worried looking faces.

Time rolled on and, both at home and at work, Jim found himself involved in detailed planning. As part of the Purchasing Team at Abbey Martin he had to run the implementation of the Purchasing Module alongside Graham and Frank as well as keeping up to speed on coordination of the other modules, which proceeded at different speeds. It was a juggling act, he reflected.

At home, things were no less hectic. Both he and Jan were constantly involved in decisions on the building process but were finding things easier now that they had sufficient information on progress being made. It had been painful to formally part company with Barry Bryson and, in fact, some of the formalities and final payments on the work he had done had still to be agreed. They were both happier that they now had the information. They were also reassured by Ian's professional approach. He delivered regular updates on all the various areas. In turn, Ian was impressed with the level of consultation that Jan and Jim were able to cope with. The walls went up and finally the roof was on. A major milestone reached.

Jan sat down to sort out the kitchen issue with Ian one Saturday morning with Jim hovering in the background and Emma – the other member of the Kitchen Team - sitting quietly to one side. 'I'm just concerned that we won't get exactly the kitchen we want. This is my one chance to get the kitchen of my dreams. I had hoped that we would be able to choose the installers, Ian, so that we could be sure of that.'

'I can understand your concern, Jan.' he replied 'but I've learned from experience that if things are to go according to plan and not encounter any serious hitches, then I need to be sure that we keep proper control by hiring only trusted contractors. Don't worry, though, you'll still get your ideal kitchen. I promise you.'

Jan looked doubtful. 'But I've read the magazines and seen lots of advertisements for lovely kitchens. I just thought that I'd be able to choose something like that when the time came. Our previous builders were going to give us a free rein.

'Ah. A free rein.' he looked serious 'I've had some experience there. A few years ago, when I was new to this business, some clients appointed a contractor who I'd not worked with before. They made a real mess of the kitchen. We put it right in the end of course but the worst of it was that the mistakes they made along the way caused serious delays to the rest of the project.' He looked straight at Jan as he continued 'And one thing everybody knows is that delays cost money.'

'I can see that.' said Jan and continued to listen with interest as Jim entered the room.

'I'll give you just one example of what could go wrong if planning is not a priority in every area. This awful kitchen contractor fitted appliances then realised they wouldn't be able to fit the tiles that had been chosen without removing those appliances. So they took them out, damaging a very expensive dishwasher in the process, laid the tiles and refitted the appliances – after they had waited six weeks for a replacement dishwasher.' Both Jan and Jim looked appalled. 'And there's more. It gets worse. They then realised that the plumbing was not correct for the appliances so they had to remove them again.'

'What about the tiles?' asked Jim.

'Yes, tiles as well.' Ian answered 'And that was just one incident in a whole catalogue of disasters. My client was not amused. The whole project was delayed for weeks, the move-in date was put back several times and the costs rose on a weekly basis. And all because the planning wasn't done properly. So now, we insist on working with contractors who know how we operate. It works.'

'We can't argue with that, can we Jan?' asked Jim, looking at his wife who he could see was disappointed.

'No, I can understand the problem. It's just that...'

Ian hurried to reassure her. 'I know, it's just that you want your dream kitchen. Our approved contractors are excellent Jan. You'll find the choice they offer is top notch. Here, let me give you their contact details so that you can choose your kitchen and put your mind at rest.' He handed her a business card.

'Thanks, Ian. I'm sorry to be a nuisance. I know that planning is important. We have to get this right. We definitely don't want to get into anything like you've just described. That sounds like a horrendous experience.'

'It was, Jan, believe me, and it's far from uncommon.'

Jim broke in. 'That could easily have happened to us, you know. Barry was definitely headed that way!'

'You're right. His lack of planning was why we had to part company.'

'There you are then.' said Ian, still looking serious 'Just choose your ideal kitchen with our contractor and we'll manage the rest.'

The very next day the Kitchen team were on the case. Jan and Emma made their initial visit to the backstreet showrooms of Ian's recommended contractor to start to plan their kitchen.

'This is a bit basic, Mum.' commented Emma as they made their way carefully through the stacked up boxes towards the small display area. It definitely did not look like the fancy showrooms they had envisaged.

Jan came to a halt in front of a row of kitchen units. She stroked the gleaming marble work surface and said 'Let's give it a chance, Emma. Some of these units are really nice.' She pulled out a drawer and commented. 'Well, that's good and solid. Feels a lot better than the kitchen we've got now.'

'Good morning. Can I help you?' The young man who had appeared from the back of the shop was casually dressed in blue jeans and a brilliant white shirt.

'I hope so.' Jan replied. 'I'm Jan Heswall and this is Emma, my daughter.' Ron Smith smiled and held out his hand. 'Ian Hemingway sent us. He's in charge of building our new house and we just wanted to look at a few kitchens. Would that be possible?'

'Pleased to meet you.' He smiled at both of them and ushered them to a small, efficient looking office space where a computer was set up and a coffee machine was gurgling and producing an enticing aroma. 'Ian told me to expect you. He's given me the dimensions of your kitchen and utility room so we need to get together a few of your ideas and requirements. Do you have an hour to spare now to do that?'

'That's very well organised.' Jan said, surprise evident on her face.

He shrugged. 'It's just how we always work. Let's sit down here. Now, tell me what you need from a kitchen. What's important to you?'

They all sat down and Ron made notes while Emma and Jan talked about what they hoped could be achieved. He started to key in some details to the computer. 'There we are. Would it look anything like that?' He turned the display screen so that they could both see it.

Jan gasped. 'That's amazing. That's just how I pictured it. What do you think Emma?'

Emma was equally impressed. Ron explained a few of the kitchen's features and tried different finishes on the kitchen units and floor but Jan and Emma decided they were happy with his first attempt. He sat back, obviously pleased with himself. 'I do like it when I get it right first time. It means that you know what you want and that I've understood you. We should be able to work well together. No problem.'

Jan relaxed. 'No problem.'

'Right, let's look at the utility area. We can get plenty more storage in there – that's so important, isn't it?'

Emma sat back, smiling. 'I'm not on the Storage Team, but I certainly think you ought to be Mr Smith.' She smiled shyly at the young man.

He grinned. 'Pete, please.'

Jim, meanwhile, was still heavily involved in the latter stages of the detailed design of the Change Project at work. This was interspersed with pre-implementation workshops about the Purchasing Module, training to bring him up to speed on all aspects of the implementation phase and various meetings about the data being gathered on the other modules. The formal sign off of the project by Neil and Tony had taken place and this was a relief to Jim who had been worried for some time that it would not get the final go-ahead. The formal review headed by Frank Littleton from the software company was the next meeting in Jim's diary.

'Morning Frank' Jim greeted Frank as they met in the corridor outside the meeting room. 'This should be interesting. I'm looking forward to this.'

'Good morning, Jim. It won't take long. It's more of a formality really. Just need to go through the modules we're working on, where we are at with each of them, that sort of thing.'

The meeting room was crowded when Frank Littleton stood up at the front of the room. 'I'll not keep you long, ladies and gentlemen, I can see that we are a little cramped in here. As you know, we're well into the detailed design phase here at Abbey Martin and implementation of three of the modules – HR, Planning and Purchasing – is due to start next week. From our point of view, these are the three easiest modules. There isn't too much configuration and they stand-alone. The processes that you have designed are very close to our recommended 'best practice' and our modules are designed along these lines. They will be a useful way for us to assess how everything is working and to offer more training where necessary. The other two modules – Finance and Order Management – need a lot more work yet. The design and configuration on those will take us perhaps another three to four months. Those modules are heavy on configuration and they share lots of dependencies.' He gave them a mountain of details on exactly what was going on with these latter modules but started to sum up when he saw a few pairs of eyes begin to glaze over. 'So, we're encouraged by the progress so far but there's an enormous amount of hard work in front of us all. Good luck with the start of the implementation. I know that it won't all go completely smoothly and we may not get all of our initial expectations satisfied but let's press on.' He sat down and he and Jim chatted as everyone filed out of the meeting. Tony soon joined them.

'I gather that went well, Frank.' he remarked as he extended his hand towards Frank and smiled at Jim in greeting. 'Have you two got some time to answer a few questions?' The two men nodded and they spent about half an hour going through the documents, ticking off the elements of the process on Frank's list for each module as they went along – design; configuration; unit testing; system testing; integration testing; user defined acceptance tests; negative testing; standard operating procedures; new documentation. 'We're ready to go with this next week then?' he beamed.

'Raring to go, Tony.' agreed Jim. 'We've another two days of training sessions to get through this week and, of course, the training will be ongoing for some time yet.

'Good stuff!' exclaimed Tony.

'Yes,' agreed Frank 'we've done extremely well to get this far so quickly.'

Jim grinned. 'I can't help thinking that something will go wrong though.'

'Careful, Jim,' Tony warned with a smile 'any more negative thoughts like that and you'll have to show yourself the red warning card!'

Jim gave himself a mental shake and resolved to keep positive. However, his negative thoughts were to be proved right even sooner than he thought. Everyone was starting to wind down for the weekend on Friday afternoon when Jim took a call from Malcolm in the IT Department. 'Are you sitting down, Jim?'

'Oh no, what is it? What's gone wrong?'

'I'll get straight to the point. The mainframe has gone down.'

'The mainframe! What's wrong with it?'

Malcolm took a deep breath. 'Nobody knows. It's completely dead. Might be something simple or it could be something that takes weeks to resolve.'

'But the implementation! It's all due to kick off on Monday. What about that?'

'At this stage all I can tell you is that everything possible will be done to resolve the problem. I'll be working straight through the weekend, obviously, alongside as many others as I can muster. But I can't make any promises until we can locate the cause of the fault.'

Jim put his head in his hands. 'This is all we need. I just knew something drastic was going to happen. Who else knows?'

'No-one yet. I thought I had better let you know first. Maybe you can break the news to anyone else who needs to know while I get on with trying to sort out the problem and arrange a repair.' His voice sounded hopeful, as though he knew Jim was the easiest one to tell and also the one he could count on to ensure that everyone was kept informed.

'Yes, yes, you're right. I'll do that. I'll let you get back to it. Just keep me informed. You've got my home number haven't you?'

'Sure. Start praying.'

After a few fraught phone calls to colleagues Jim headed home, satisfied that he could do no more to help and hoping for a miracle over the weekend.

On Sunday, Jim and his family welcomed John and Brenda to their house for a barbecue. It was a beautiful day and when the older couple arrived, all the Heswall's were already outside, enjoying the sunshine. When the drinks had been dispensed, Jan showed Brenda her favourite plants.

'You're going to miss this, you know Jan.' Brenda looked around her admiringly while Jan reflected on her garden with pride. It was all her own work.

'I've lived here since before the girls were born so it's evolved gradually. Lots of hard work though.'

'I can imagine. In a couple of months you'll be living in what feels like a building site – no plants, nowhere to entertain outside.'

Jan laughed. 'But that's one of the things that I'm looking forward to. I can't wait to start again. I can avoid all the mistakes I've made with this garden. I've got lots of plans for the new one.' She looked at Brenda's doubtful face. 'Come on. Speaking of plans, I'll show you the plans for the new house. You'll love it.' She headed for the house, with Brenda following.

'Here's the kitchen that Emma and I chose this week. The designer was quite dishy. He really knew his stuff though. We've got lots of storage, all the appliances you could wish for, lovely work surfaces – the lot.'

Brenda could not help smiling at Jan's enthusiasm but felt compelled to try to warn her of the hard times to come. 'Sounds lovely but you know it won't be all plain sailing, don't you?'

Jan brushed the warning aside. 'Oh, I know that. But won't it be wonderful? A new kitchen, lots more space, a garden to plan and develop all over again – but this time I will know what I'm doing.'

'But the work!' exclaimed Brenda 'And things can always go wrong, you know. I don't want to sound too pessimistic but you know what they say about the 'best laid plans...' This whole project could be really painful.'

Jan leaned against the kitchen counter where she had spread out the house plans and kitchen drawings while Brenda had been issuing her dire warnings. 'I know things don't always go smoothly but Brenda; it's just so exciting. I can't wait for it all to be ready.' The enthusiasm shone in her bright blue eyes and Brenda decided that there was no point in saying more.

The two men, meanwhile, were presiding over the barbecue and having a similar conversation. 'How are you two coping with the pressure of building a house?' John enquired as casually as he could while sipping his beer.

'Pressure? I wouldn't exactly call it pressure. We're all looking forward to seeing the finished result.'

'But it will be a painful process, you know.'

'No, it's something we really want. And something we desperately need.'

'That may be so, but it can still hurt.' John picked up a barbecue fork and prodded the steaks and sausages while he considered how far he ought to go down this line of argument. 'There's the finances to think about, the hassle of dealing with builders and don't underestimate the pain for Jan and the girls of moving out of their home.'

Jim took the fork. 'These are going to burn.' He deftly moved the sausages to a plate held out by John. 'But they smell good, don't they?'

John smiled in agreement but said nothing, waiting for Jim to return to the subject of the house move.

'I know it won't all be easy but it's definitely a move for the better. Let's not get too serious about it. We've enough problems to worry about at work. The mainframe problem was just about the last thing we needed at this stage. I wonder how that's going. I haven't heard anything so far, have you?'

John shook his head. 'It's not looking good for tomorrow, that's for sure.'

Taking a deep breath Jim turned back to the steaks. 'Nothing we can do about it right now. This is the last barbecue we'll have here so let's make the most of it.'

Points to reflect on

- Evaluate the risk in your transition and refine as needed
- There will be pain – it is unavoidable. Manage expectations
- It is not 'red card behaviour' to talk through potential problems and set contingencies
- Constantly evaluate the issues that cause pain in the project and attempt to minimise
- Ensure that executive management is visible and supportive in the painful decisions

Chapter 18 – Detailed Design Part 2

As soon as Jim saw the IT Manager's face on Monday morning, he knew that he did not have good news. 'What can you tell me, Malcolm? Is there any good news at all?' he asked, the resignation obvious in his voice.

'I'm sorry Jim. There won't be any implementation today. The whole business is at a standstill, I'm afraid. We don't even really understand what the problem is. We've had the hardware manufacturers on site for the last twenty-four hours. It's a nightmare.'

'You're telling me. Any estimates of when we'll be up and running again?'

'Not really. We are doing our best.'

'Fine.' said Jim, shortly, 'I'll call a Steering Group meeting right away. Just keep me informed.'

John was just going into another meeting when Jim called, but Jim managed to get the others together within an hour and they went ahead without John. They were sitting around the meeting room table, looking glum, when Malcolm burst in. 'Good news. We should be up and running by tomorrow morning.' They all breathed a collective sigh of relief. Malcolm continued 'As usual it was something quite simple – once we'd found the fault. We're reprogramming now. We need to know when will you go live with the new software.' He looked first at Jim then at Frank Littleton.

'Give us a few minutes Malcolm, we'll reformulate our schedule and I'll give you a call.' Malcolm backed out of the room and everyone started talking at once. Jim held up his hands. 'Hold on there. Let's just look at when is the earliest feasible date to go with the implementation of the first two modules.' He checked his diary. 'Frank, what do you think – can we go for it this week?'

Frank hesitated, scratching his head, but came up with a decisive answer. 'Yes, let's reschedule

Wednesday's work and go for it then. We were ready on Friday last week until this happened so I don't see why we can't be ready within forty eight hours.' He slammed his pen down on the table and looked around, challenging anyone to disagree. But no one did.

Jim beamed. 'Wednesday it is then.'

Jim reflected on the day so far as he strode across the warehouse that afternoon. He had gone from near despair to cheerful optimism. Little did he know that he was just about to swing right back again. He passed several warehouse operatives, muttering in a corner. They stopped as he squeezed past between the group of men and several cartons of product that were stacked up just outside the office. Jim's irritation increased as he saw Ron's handwritten stock figures on the cartons – another problem that had not gone away. He went into Alison's small, dark office and she looked up, surprised to see him. 'What's the matter with that lot out there?' asked Jim.

Who? What do you mean?' she seemed distracted but Jim pushed on.

'Those guys out there. They seem a bit unhappy.'

Understanding dawned on her face. 'Oh, them. They're always unhappy. There's been a bit of gossip about the figures out last week. It started on Friday. They seem to think that the Change Project is to blame for everything around here. If the coffee machine broke, they would blame it on ERP!' she said, trying to lighten up the conversation but not succeeding. He looked troubled.

'What is it this time then?' he asked

'You know – the poor sales figures and also the rumour that is spreading that the computer blew up because it was overloaded with the new software.'

'For heaven's sake, that's ridiculous!' he exploded.

'Don't shoot the messenger – you did ask!' she protested.

'Sorry. I've a lot on my mind.'

'Yes, I can imagine.'

'Anyway, I'm interrupting you. What's that you're working on?'

'I'm just trying to catch up with a few things while the mainframe is down. It will be chaos around here when it comes back on line.'

'You're right. I'll not take up too much time. I just need to check a few things for the newsletter, then I'll get out of your way.'

He spent the remainder of the day gearing up everyone in Purchasing and HR for the implementation now starting on Wednesday.

When he arrived home he found Jan in the kitchen on her knees. 'What are you up to?' he asked.

Her muffled reply came from the depths of the crockery cupboard. 'Just trying to clear a few things out ready for the move. I've got rid of things I didn't even know we had. We've got so much junk.'

'I can see that.' he declared picking up a teapot in the shape of a scene from Goldilocks and the Three Bears. 'Where on earth did you get this thing?'

'A present from my mother. Couldn't you guess?' she answered as she struggled to her feet clutching a pile of mismatched saucers and kissed him quickly on his cheek. 'I don't really know what to do with it. I'm sure she would hate it if she knew that I'd thrown it away.'

'Will you ever use it in this new dream kitchen of ours?' he asked incredulously.

She grimaced and shook her head. 'No way.' She giggled as he turned it over in his hands. 'It really is grotesque isn't it?'

'It certainly is. It doesn't have a place in our new life then does it? Give it to charity or something. Just don't take it with us. The new streamlined system doesn't need it.' He was thinking about all the old systems that he had to make sure were dispensed with at Abbey Martin – the manual logbooks, the inch-thick diaries, the whiteboards on the office walls, the spreadsheets designed by individuals tracking data that would repeat the work of the new integrated system. He saw the discomfort and indecision on his wife's face. 'Please, love, just get rid of it. If it's been at the back of the cupboard for all these months, your mother will never know. She's probably forgotten all about it.'

Jan nodded but Jim noticed that she placed the teapot on the table, rather than in one of the boxes of junk waiting by the back door.

He sighed and asked 'What's for dinner then?'

'There's a casserole warming in the oven. Go and round up the boys. The girls won't be back for another hour and they've already eaten. It will be ready in ten or fifteen minutes.' She looked subdued as she started stacking things back in the cupboards.

Over dinner the boys alternated between giggling and arguing while Jan and Jim discussed how the building was going. 'Don't forget we've got Ian Hemingway coming over on Saturday morning to give us an update.' reminded Jan 'He's pretty good at that, don't you think?'

'Yes, he certainly keeps us informed. Didn't he say we'd be looking at the detailed budget in particular this time?'

'Mmm...'

'You're not worried about it are you?' Jim looked at her as she moved the food around on her plate and he became aware that she had not eaten very much of her meal. He and the boys had cleared their plates several minutes ago and the boys seemed to be waiting – a little impatiently – to see if there was any dessert.

'Get a banana each boys. I haven't had time to make anything else.'

James and Andrew said nothing as they scrambled off their chairs and grabbed bananas from the fruit bowl on their way up to their bedroom. Jim decided not to call them back to clear the table but let them go without comment. He needed to talk to Jan and find out exactly what was worrying her. 'OK. What's the matter?'

Jan studied her husband's face. 'Aren't you worried about the budget or the move? Have we done the right thing? Perhaps John and Brenda were right. This might be too much for us to handle. Perhaps we should stay where we are.' She put her head in her hands and Jim leaned across and lifted her chin so that he could look directly into her eyes.

'Hey, this isn't like you. I'm depending on you to be positive.'

She burst into tears. He rushed around the table and held her until she stopped, taking great gulps of breath. 'I'm sorry. I think I'm just tired. That stupid teapot set me off.' She half laughed, half cried.

'We've never even used that teapot. It can't be that important!' he said.

'No it isn't important at all. It's just that I've just realised that we'll be leaving everything behind here. Perhaps that's what was upsetting Philippa a few months ago. Perhaps she realised before I did, that this marks a real change in our lives.'

Jim exclaimed 'Of course it's a big change in our lives. It's a change for the better. You want a new home don't you?'

'Of course I do but maybe this is what Brenda meant about it being a painful experience.'

They talked and talked, reassuring each other that they were doing the right thing and that, anyway, it was too late now to go back. The boys crept in to the dining room and then crept out again, realising, for once, that they should make themselves scarce. An hour went past and the girls came in. They helped themselves to a snack while Jan and Jim continued to talk. There was an obvious worry that they would have financial problems but they agreed to face that after Ian's update visit. In the meantime, Jan came back around to Jim's way of thinking – that there was more good than bad in the change they were making.

The following day Jim was involved in pre-implementation training on how all the parts of the new system would be interlinked and how this would make people's jobs easier. He had found this whole concept difficult to deal with and to fully comprehend. To help him and various other people in the business, Graham had put together some 'storyboards'; which showed how processes would work post-ERP. These used actual work situations and showed how the new systems would help them to work more efficiently and achieve more than was possible with the present systems' drawbacks and inefficiencies. He had decided that these were to form the basis of his presentation today so he felt well prepared and he was now getting used to his role in training and motivation.

He entered the training room. 'Good morning, everyone. Let's get the ball rolling. I know some of you may be feeling a little uncertain about what is going to happen in the next few weeks, so let me put your minds at rest.' He went on to run a slick opening session aimed specifically at motivating the system's doubters. Graham took over after the coffee break, concentrating on the more detailed aspects of how the system would work on a day-to-day basis.

As usual after a major meeting or training session, Jim reported back to Tony on the progress – or otherwise – that had been made. There were the searching questions that Jim had got used to over the last few months and Jim marvelled yet again about Tony's never-flagging enthusiasm and command of the level of detail involved.

Not everything went quite so smoothly that day however. As Jim came out of the session with Tony, Frank was waiting for him with Graham. As soon as he saw Frank's face, Jim knew that there were more problems to be dealt with. What's up?' he asked.

Both started to answer at once but Graham gave way to Frank. 'Last minute configuration problems.' stated

Frank baldly 'Nothing I haven't seen before but it could hold things up a bit tomorrow.'

'I'm not sure I want to hear this.' said Jim, near-despair crackling in his voice. 'This is just too much. We're going live with two modules and, let me get this right, you're telling me that the system isn't set up correctly now?' He glared at Frank and Graham who both looked down at their feet.

Graham spoke first. 'Don't jump to conclusions, Jim. It's not that bad. Just a last minute hitch that we can get sorted out.'

'I see. Perhaps you'd better explain.' Jim tried hard to contain his anger and disappointment.

Frank took a breath. 'Well, the workflow routing isn't working correctly. It's relatively simple to fix but the implication is that Purchase Orders don't get authorised. Potentially therefore raw materials or new goods don't get ordered or they're delayed. The knock on effect of this, of course,' he said glancing quickly at Jim to make sure that he was still following his brief explanation.

As the software expert paused, Jim voiced his thoughts aloud 'That's just not a risk we should take, is it? If that happens there's ultimately a probability that our customers will suffer then our sales suffer and profits go down.'

'Exactly. It's fairly simple to fix but it's a large scale issue and we can't carry on until it's fixed.' said Frank, relieved that Jim could see the consequences.

'Is there anything I can do to help? We want to minimise any delay.' Jim asked.

'That goes without saying.' said Frank 'We're working as fast as we can and no, I don't think there's anything you can do. We'll handle it. We just wanted to make sure that you knew.'

'I suppose you want me to keep Tony in the picture?'

'That might be helpful.'

'OK, but you shouldn't think that I will be able to keep him off your backs. He'll want all the details and the implications, you know.'

'Tell us something we don't know.' whispered Graham and then, in a louder voice, 'Tell him I'll give him the full story as soon as the emergency is over. Thanks, Jim.'

'Right then. I'll see you tomorrow – hopefully bright and early for the launch of the new systems.'

Jim drove home slowly that evening, feeling the weight of responsibility on him both at home and at work and promising himself that he and Jan would have at least a weekend away as soon as all this madness was over and they were ensconced in their new home.

The day of the first implementation and the 'live' system came around and Jim sat in his office, awaiting the onslaught of queries and problems that he expected the 'switch over' to bring. But all was quiet. Too quiet, thought Jim so he wandered over to find Graham. He discovered the consultant, together with Frank Littleton, in the meeting room, surrounded by configuration and planning documents. 'You two look busy.' started Jim 'Plenty of problems and feedback coming in, then?'

Graham looked up. 'No, very little actually. But that's only to be expected. If this goes the way of all the other implementations that I've been involved in or heard about, I think you'll find that some of the issues will start to come out a week or so after going live. Where the training hasn't quite sunk in – that sort of thing. Others will build over time but they will show up in my exception reports.'

'Right, I can look forward to that then!' Jim's worries showed in his unusually sarcastic tone but he still took the opportunity to learn a bit more from Graham's obvious experience. 'What sort of issues will they be then?'

'There will be people issues – you might have to deal with some of those. The Helpdesk we've set up will field most of the questions on using the system that will come up. I just hope there won't be too many problems with the configuration.'

'What do you mean?'

'Well, we've done a lot of testing to make sure that the system works but it has mostly been positive testing – trials that use data that assume the system works and that all the data that is input will be right. I think maybe we needed a bit more negative testing but Frank is in charge of that. We'll see.' He glanced at Frank who was doodling on his pad, obviously preferring to stay quiet on the issue of testing.

Jim looked puzzled and no less worried than when he had started this conversation. 'I'll leave you two to your work then and get on with something else.'

Late morning, Tony came into his office. Jim stood up in surprise. 'Tony! How are you? I didn't know you were working here this morning.'

'I've been working with Neil.' He sat down and slumped in the old chair that Jim kept in his office for what was usually only the occasional visitor. Perhaps he ought to requisition a better chair. He could try out the new purchasing system, he mused.

Tony straightened up. 'We've been going through the sales figures for last month. They don't make happy reading. Of course, Neil is worried. Says we've 'taken our eye off the ball' by concentrating on the ERP project.'

Jim looked aghast. 'That's just ridiculous. The sales figures are nothing to do with the Project.'

'You try telling Neil that. His Sales Manager has already been bending his ear, giving him the idea that all the sales staff are worried about the changes and haven't been able to get the help they need because everyone is busy with the project.'

'Is there any truth at all in that?'

'No, I'm certainly not thinking that way, but the project is a useful scapegoat for anyone not one hundred percent behind it. I'm a bit surprised that this attitude of blaming the project has started so soon, I have to admit.'

'So, what will you have to do?' Jim saw an unusually serious look on Tony's face.

'Well, we've already scheduled an extra review of costs. We've had some savings and some real benefits – such as the reduction in accidents on the shop floor and the number of customer complaints is down and so on. I suppose we've got a morale problem just now but that's only to be expected at this stage in the game but sometimes people can't see that this sort of thing is only temporary but it will impact on sales and they just blindly blame ERP. Getting everyone working in the same direction is the really difficult bit. People problems can't be allowed to derail the project.'

'Anything I can do to help?' asked Jim.

'Just keep spreading the word, Jim, that's all.'

'Good. How about the implementations? How are they going?'

'Just a few minor problems about authority level setting, that sort of thing, but we're getting there.'

They discussed the implementation of the Purchasing, Planning and HR modules in detail until Tony, satisfied with what he heard and seemingly fortified by the more positive approach he found in Jim's office, left to return to his offices. Jim was glad that he had not divulged his observations about Frank and Graham's worries.

Another Saturday morning; another meeting with builders. Ian had laid his papers out on the dining room table and Jim and Jan were waiting with mixed feelings. They were still excited by the new house and still convinced that all would be well. However, Ian had warned them of two potential problem areas. Also, Brenda and John had been adamant that the experience could be painful so, despite their extreme optimism, they were conscious of a small inner voice trying to tell them that problems lay ahead.

Ian cleared his throat. 'Let me just run through some of the changes in scope over the last few weeks.' He began to count on his fingers 'Firstly, we've added in a path around the conservatory. Top grade materials – you'll have to continue using those in the rest of the garden, of course, when you get to that stage but I think you've made a first class choice there and doing it now will minimise the disruption later. Secondly, the pool definitely can't be done just now. As we discussed last week, the budget won't stand it right now, but it's something to look forward to for the future. Any comments there?' He looked expectantly at Jan and Jim.

Jan returned his stare impatiently. 'We know about those things. What we need to know is how they affect the total cost and the timings.'

'Give him a chance, Jan.' She glared at Jim and fidgeted with her blue book but kept quiet. 'Carry on, Ian.'

'Thirdly, we've arranged the special materials to comply with the environmental concerns you raised at the beginning and the costings for the kitchen are in. That brings me to the fourth concern.' He produced a chart from the pile of documents on the table. 'This is a Gant chart – it shows when everything is scheduled to happen, all the important milestones and the phasing of all the work. It's bang up to date. I put all the new information into it last night – the materials delivery dates, the new garden path and so on.'

Jim leaned forward for a better view. 'Graham uses these at work.' He saw the enquiring look from the young architect and explained 'Graham is a consultant who's dealing with our Business Change Project.'

'I see.' said Ian 'The important information from your point of view is the Completion Date when we hand over the property to you. That would allow you to set your moving in date and to schedule the sale of your existing house thereby releasing money to pay for this job.' Both Jim and Jan leaned forward eagerly. 'I'm afraid, as you can see, that we've had to extend the time by four weeks at this stage. That takes into account the extra time for the delivery of the special materials. We've managed to phase the extra work in the garden in so that it doesn't put even more time onto the project.'

'Four weeks. That's not too bad. Is it, Jan?' Jim kept up his customary cheerfulness and optimism but Jan looked devastated.

'I was still hoping we could improve on the moving-in date, not make it worse.' she replied.

Ian hurried on. 'The other important information is here, on the financials sheet.' He fished out yet another piece of paper. 'There's just no way we will keep to budget, we're looking at a three percent overspend. It was inevitable, I suppose.'

'Yes.' agreed Jim 'Inevitable, but that doesn't help our finances.'

'Don't forget that you will probably benefit slightly from the rise in house prices at the moment – it's a buoyant market so that will help.'

Ian drained his coffee cup and Jan was so distracted that she did not even notice. He got to his feet and gathered his paperwork together, rolling up the plans and putting them carefully into a large cardboard tube.

After showing Ian out, Jim came back to the table and Jan looked up at her husband. 'The extra four weeks is the least of our problems, isn't it?'

'Yes. The budget is the real headache.'

They both sat at the table for some minutes, not speaking, just thinking – and worrying.

Points to reflect on

- As stated so many times - active sponsorship is fundamental
- ERP will probably be blamed for almost anything that goes wrong, with little recall of problems prior to the project
- Build in extra time and budget. There will always be the need for the unexpected
- If people do not 'buy in' to the project, it will result in extra cost. Re-communicate the benefits
- Be careful not to change decisions and move forward. If you change decisions you need to re-evaluate benefits and risks

Chapter 19 - Implementation

The weekend passed and Jim and Jan were still worrying about the over spend on their budget. It brought them to the absolute limit of the money available to them. Their savings had all but gone and they were faced with the daunting prospect of a visit to the bank.

When they woke on Monday morning. The decision had been made. 'We don't really have any choice, Jan' admitted Jim over breakfast 'so if you can find time to set up an appointment with the Personal Banker, we'll go together. Try for an afternoon towards the end of this week or the beginning of next.'

Jan sighed 'We've borrowed to our limit already. Do you think we'll be able to pull this off?'

Jim pulled her to him, smiled and said, as positively as he could manage 'We will. It might be difficult, but we'll do it. You're earning a bit more with the extra shifts you're doing and when we are in the new place our running costs will be less – think of all that environmentally friendly insulation that Emma was so keen on. Don't worry. It's too late to go back now so we just have to keep going forward.'

'You know, paying back what we will have to borrow will be even more difficult now. This whole thing is getting so stressful, it's untrue.' She moved away from Jim saying 'Still, worrying about it won't get us both to work on time and the kids off to school.' She went slowly up the stairs to sort out the children. 'Come on, you lot. I've no time or patience for messing around this morning.' All four children filed down the stairs in various states of readiness. All were aware of the tense atmosphere there had been in the house over the weekend and knew better than to play Jan up. Jim looked on, worried that the stress was getting to Jan. She certainly wasn't her usual sunny self this morning.

However, he did not have time to hang around and he hurried off to work with the budget problems at home going round and round in his mind. He was unable to reach any other conclusion than that they had to enlist the bank's help to carry on.

Things did not lighten up when he reached the factory either. He was called straight into a meeting with Graham and Frank. He went into the meeting room and saw that John was already there. 'What's going on?' he asked, even before sitting down or helping himself to coffee.

'It's the budget.' stated Graham in a matter of fact way 'We will definitely be over budget. Frank's been working on the figures for the next two modules over the weekend. The costs were underestimated I think.'

'Could you go through the details for us Frank?' he looked at the young man.

Frank cleared his throat nervously. 'Yes, sure.' He picked up a few sheets of papers covered in figures and Jim thought he detected a slight tremor in Frank's hands but his voice was steady and he sounded as knowledgeable as ever. 'As you all know, the next two modules are the most intricate and will also have the largest impact on the cost savings that we know are possible from this ERP implementation. We're getting to the end of the Data Gathering and Detailed Design phases now and it has become obvious that some areas of these modules require extensive customisation. In particular, the Credit Control area of the Finance Module and the Stock Availability details for the Order Management Module have been subject to a lot of customisation. The extra work and expenditure will show dividends in the end of course, but it does mean a revision of the overall budget.' He looked up and met the eyes of the other three men.

Graham took over again. 'We need the board's approval for a budget increase like this.'

'Getting approval will be hard – if not impossible – while we've got this downturn in sales, won't it?' Jim looked directly at John who, as Tony's 'right-hand man' in Finance at Abbey was, of those around the table, the one most likely to know the answer to that question.

John frowned. 'We will have some explaining to do and we'll have to do a hard sell on the benefits again but I think we should be able to swing it. We are a long way down this road, after all. It would be commercial suicide to stop now.'

'We don't have much choice, do we?' asked Jim, thoroughly fed up with this lack of choice on all fronts.

Back home in the evening, the stress levels continued to rise. The boys were fractious, picking up on the edginess of the adults and the girls were stroppy in response. Yet another evening was spent going over and over the budget and neither Jim nor Jan could make the figures work. Jan had managed to get a late afternoon appointment with their Bank Manager scheduled for Thursday. They turned their attention from trying to find ways to economise and trying to cut any excess spending in their budget to putting together a case for borrowing more as a short-term measure. Jim tried not to let Jan see just how worried he was. He reasoned that the stress levels at work were probably making him feel even more nervous of the changes at home but resolved just to try to work through both of the change situations as best he could. He wanted to get through without unduly bothering either his wife and family or his colleagues at work.

Before he could try out his persuasive powers on his Personal Banker, he had to deal with Tony. The Managing Director had, of course, been informed of the shortfall in the budget and had called an immediate meeting so that he had the opportunity to ask questions of those closest to the project.

'Graham, perhaps you can give me a bit of background information. Where are we at with the implementation of the first two modules?'

The consultant looked unblinkingly back at Tony and answered 'Running like clockwork.' He folded his arms and waited.

'I take it you are pleased with the first part of the implementation then, but give me just a bit more detail Graham.' Tony tapped his pen on the table and glanced sarcastically at Graham.

'Sorry, Tony.' He proceeded to give a detailed rundown of the progress so far and was rewarded with just a short nod from Tony who then turned his attention to Frank.

'Progress on the remaining modules, please, Frank.' His request was brief and Frank knew exactly what was required. He updated Tony on the current situation then moved swiftly on to the budget problems with the Order Management and Finance Modules.

As Frank came to the end of his summary, Graham took up the case. 'We ought to take this opportunity to review some of the benefits. There are two main areas; payroll savings and reduced costs in the Purchasing area.'

'Have you got the figures handy, Graham?' asked Tony. Jim smiled to himself as Graham flashed the appropriate details on his laptop screen – when doesn't this man have the relevant figures handy? he mused but kept his attention on Graham.

'The release of staff from 'merged functions'.' He pointed to his screen. 'The combined sales force – we've saved two manager's salaries there plus the representatives where there was overlap, big savings in the finance area – credit controllers etc - and in HR.' He raced through all the areas where staff reductions had already been made, quantifying the savings as he went through the figures on his screen. 'Then there are the savings in the purchasing department. We've got savings coming through from better discounts and single supplier deals that have already been done. All that gives us a sizeable sum to cover some of the costs incurred and to come.'

Tony thought for just a few seconds after Graham and Frank had finished, then summed things up so clearly that even Jim was glad that he was in on this meeting, difficult though it was. He now knew just where the increases in expenditure had come from but he also had a clearer picture of the benefits that were being gained form the project. He almost held his breath as Tony considered the position. 'OK we go with the extra expenditure. But I want a full report from you Frank on the extra software costs and from you Graham on any overspend on your costs. We'll discuss anything above this current level on a weekly basis from now on.' His expression left none of them in any doubt as to the seriousness of what had just been agreed and they all heaved huge sighs of relief as he marched out of the meeting room.

Graham, Jim and Frank remained in their seats for some minutes after Tony's departure. Graham broke the silence. 'That's all we need – Tony laying down the law. We've enough problems with all the changes to operating procedures that we're trying to push through.

'Not to mention the training and staff changes.' grumbled Jim.

'Or the system configuration problems.' Frank chipped in.

Graham put his face in his hands and the others struggled to make out what he was trying to say. 'Hey, listen to us. We need to concentrate on the good stuff. The cost benefits are just around the corner. I think we need to sell this project all over again.'

'Who to? Ourselves?' said Jim, grinning broadly.

Graham looked up. 'That might be one of your better ideas, Jim. First we'll convince ourselves that we're getting to where we want to be, then we'll work on

Tony. Let's look at the figures again, Frank.' All three got down to work. After two solid hours of work, Jim was again nominated to speak to Tony and a meeting was quickly arranged.

'So you see, Tony,' he concluded 'Graham has already shown you the savings from the work we've done on the Purchasing Module. We'll do similar work on rationalising customers shortly after the implementation of the Order Management Module. The rationalisation will drive changes in warehousing, transport, more savings in the customer service and credit control departments as all the changes take effect. We're driving costs out the whole time so the projected savings are looking good. And, of course, we've planned in weekly updates on the budget – I'll be taking care of those for you.' He stopped, holding his breath as he looked at Tony and suspending logical thought as Tony returned his stare.

'You're right, Jim. You're spot on. We just need to keep going until the real benefits come through. That's what we started this for, after all.'

As Jim walked briskly down the corridor back to his own office, his step was considerably lighter than when he had been on his way to see Tony. He knew that keeping the end result in mind was the way to keep going. This was, of course, equally true for the changes they were trying hard to make at home. He must remember to remind Jan tonight – certainly before the meeting with the Bank Manager on Thursday

On Thursday, however, Jim did not find it easy to get away from the situations at work – there seemed to be a new crisis every couple of hours and he rushed through the front doors of the bank at precisely ten minutes past four. He cast about looking for someone who could help him.

'My wife and I have an appointment at 4 o'clock with Mr Kramer, my wife is meeting me here.'

The slim, smartly suited receptionist gave him a cursory glance. 'And you are?'
'Heswall, Jim Heswall.' Jim announced in flustered tones.
'Ah, Mr Heswall. Yes,' responded the young woman in a way that to Jim suggested disapproval 'your wife is in with Mr Kramer now. She has been here for about twenty minutes. Follow me.' Jim knew this woman was used to being in charge in her little domain and he dare not argue. Instead he followed quietly behind as she led the way into a plush office at the rear of the bank.
Harry Kramer stood up as Jim was shown into his office but Jan remained seated. She glared at Jim but he did not meet her stare as he apologised to the Bank Manager for his lateness. His voice tailed off as he mumbled his excuses '...lots going on at work...'
'As I was just saying to your wife, Mr Heswall, we need to go through your budget with a fine toothcomb. She's already shown me your Architect's latest update and, I must say, everything seems in order there. He ran his perfectly manicured index finger down the figures written in Jan's neat handwriting. 'Yes, here we are. Could you just run through these figures for me Mrs Heswall?' Jan leant forward while Jim leant back in his chair. He was aware that he was being left out of the detailed discussions but knew that Jan had a real grasp of the figures. By the time she had finished with the detailed explanations and they had both assured the Bank Manager that their jobs were as reasonably secure as could be expected in the current climate, Harry Kramer was ready to sign off the latest stage in their extensive loan process suitably enhanced to cover the new requirements. The pressure was off – for the time being.
The pressure was soon reapplied when they got back home and found an urgent message on their answer machine from Ian Hemingway. Jan rang him back

straight away. Jim could only hear one side of the conversation but what he did hear was not comforting. 'Two more weeks? But it's less than a week since you told us that the handover date was sorted out!' Jan's voice rose until she was almost screeching into the phone. 'But you can't be serious. You can't do this. We've made arrangements, ordered furniture and so on.' She listened, her face getting pinker by the second. 'I see. Well, we'll see you on Saturday morning then. Goodbye.' She replaced the handset with a bang and burst into tears.

Jan was inconsolable, almost hysterical. No matter what Jim did or said, he could not get her to talk sensibly about the necessary changes to their plans. Her parents had planned to visit immediately after the move and had already bought their tickets to coincide with the children's holidays from school. They had new furniture on order – the king-sized dining table that Jan had had her eye on for weeks at the designer store in the city – and a specially made set of furniture for their bedroom. He gave up trying to make her discuss the problem and settled on ordering pizza for supper. Jan was in no fit state to cook. Between sitting glassy-eyed in the dining room staring at her budgets and timetables, asking in a shrill voice where they could possibly put all the extra furniture, and frantic calls to her mother, she had no time or energy for reasonable discussion and rearranged plans. She even refused to talk to the girls when they asked whether they were still going to be going shopping on Saturday to choose new carpets.

The following day Jim could see signs that she had picked herself up and was back to her usual fighting self. He held her as he said 'That's more like it. I'm just used to you getting on with it. Whatever happens, we'll manage.'

Jan was not quite back to normal. Tears welled up again. 'It feels like we will never get into our home. It's costing a fortune, the children are arguing non-stop, we can't seem to settle on a handover date, the furniture will be delivered to a building site, my mum and Dad will lose the money on their tickets...' she paused for breath and looked hopelessly at Jim.

'Slow down. Let's deal with one thing at once. Make one of your lists.' He sat her down at the table and put her blue book and a pen in front of her. 'First, that furniture. Can't we ring the shops and ask them to store it? So long as they've got their money, they won't worry about an extra week or two in the warehouse.'

Jan did not look convinced. 'They might sell it to someone else.'

'Not if we pay them first, they won't. I'll ring them from work tomorrow.' He just hoped he would be able to find time.

Jan got into the swing of the job and quickly checked the items on her list. 'That just leaves Mum and Dad and their tickets – they will lose their money, I think.'

'Well, we can't do anything about that. I'm sure they will understand and they will be able to sort something else out. If the worst comes to the worst, they'll have to visit us here instead of at the new house. They could help us to pack everything, couldn't they?'

Jan managed a tiny smile. 'Hey you! Looking on the bright side is my job.'

'Oh no it isn't. You've been promoted to re-scheduler. Get on with your work!'

By the end of the following day, Jan and Jim had rescheduled all the arrangements they had made when they had thought that the move date was fixed. Jim's in-laws were to visit as part of a leaving party. They had promised to bring their oldest clothes so that they could help out and were already planning a visit a few months later to see the new home. Disappointment and stress were still a problem but yet another crisis had been worked through. As Jim explained to a mesmerised Jan (or maybe she was just bored, reflected Jim) 'Dependencies dictate the sequence of events', that's what Graham always says when we have to change anything.' She shrugged her shoulders and left him to his musings and to her retreating back he muttered 'No, I'm not sure what it means either – just do things in the right order, I guess.' He stopped and ran his hands through his hair. I must stop talking to myself, he thought, that must be because I'm so tired. No wonder I'm going mad with the hours I'm putting in – both at home and at work. Still, the end is in sight, I think...

Points to reflect on

- Expect delays – unexpected events should be expected – build contingency into the plan
- Emotions will run high as deadlines approach
- Benefits need to come through from process changes – not just from software
- Quick wins need to be found and implemented – providing the 'fit' with the end solutions. These 'quick wins' will also boost positive morale
- Stop any holdover 'as is' processes. Ensure the 'to be' model is in place
- The first day will be scary. Get executive management walking around
- Manage the issues through a disciplined approach. Take out emotion – the 'squeaky wheels' will emerge and demand undue influence

Chapter 20 - Changes

The following week at Abbey Martin was the first complete week after the Purchasing and Human Resources systems had gone live and the changes kicked in. The Helpdesk was fielding as many problems as they could cope with. At the end of the week, as had become usual practice since the tense meeting to increase the budget, Jim was giving Tony an update of the progress that week.

'Sixty-three percent of the calls to the Helpdesk have been about learning issues. To put it simply – they've been about people forgetting what they have been told. Mostly just forgetting their system passwords. Everyone was e-mailed with their password, of course, but so many people just took no notice...' He looked up at Tony's disbelieving face 'Yes, I know, it's hard to believe but Graham tells me that's exactly what he would have expected at this stage. I've asked him how we can get round it when we go live with the major modules but he says there is no solution. We just have to take the pain.'

'That's ridiculous! That's just poor management. We must stamp on that one before the next stage of implementation. Jim, you should make sure all the departmental managers are aware of this potential for forgetfulness. Get them to take some responsibility for making sure their people are up to speed. What other problems?'

Jim nodded. 'People forgetting how to enter data and so on. I think people just freeze up when faced with something new. We just have to keep reinforcing the training. We've also had two people complaining that they've not got enough to do.'

'What?' annoyance flashed on Tony's face again.

'Apparently this is normal too. I've gone through an analysis of the Helpdesk calls with Graham and Frank.'

'Well, it isn't going to become normal in my company.' said Tony.
'No, no, it just highlights where we can sort out some redeployment. It will all shake down if we give it a couple of weeks.' said Jim hastily. 'Perhaps more worrying is that a series of orders failed. We had to confirm them by fax to make sure they went through. It was a regular supplier so no real harm was done. Frank is working with the Purchasing Department right now to sort it out.'
Tony looked at his watch. 'It's a bit late for a Friday, isn't it?'
'It's important to get it right. Everyone has to be prepared to do a bit extra – not just us you know.' He grinned at Tony and stood up. 'The budget figures are unchanged this week so unless there's anything else you need from me, I'll get over to Purchasing and check on Frank's progress.'
'You look tired, Jim. I think we will all need a bit of time off when this is up and running.' He smiled as Jim left the boardroom.
The weekend at home passed in a blur of packing and Jim realised that work was taking up so much of his time and of his thoughts that he was not as involved in the change process at home as perhaps he should be. Jan was handling most things alone and the strain was showing on her. She was irritated with the kids and quick to show it when they did not do exactly as they were told. She seemed so caught up in the building project and the arrangements for the move that she had no time for Jim's news or his moans about the people problems at work. By late Saturday afternoon, Jim was looking after the boys and sitting morosely in front of the television with a beer, feeling sorry for himself. Jan was out shopping with the girls – looking at bits and pieces for the new house, he thought. In truth, he missed her. His wife was usually the one who jollied him along, kept the whole family moving forwards but she seemed to have just put her

head down and decided to get on with the entire job herself. Perhaps she even thought that she did not need his help, he reflected. The boys interrupted this introspection when they came in from the garden and begged to be allowed to play with a game that had been packed away in a large carton in their bedroom.

Andrew hopped from foot to foot as he pleaded with Jim. 'We know where it is, Dad. Can't we get it out and play with it? We promise we'll put it away again.'

Yes, Dad, we'll put it away.' James chipped in.

Distracted as he was with his thoughts of problems at work and feeling aggrieved at the apparent neglect by his wife, Jim could see no harm in letting them have their way and gave his permission while taking another beer from the refrigerator.

'Where are the boys?' asked Jan as she walked through the front door and saw Jim sitting watching television.

'Upstairs, playing with a game. They have been quiet for a while.'

'Right. I'll check on them when I've changed in to my jeans ready to do a bit more packing and junk sorting.

She went slowly up the stairs. Within two minutes she was down again and facing Jim in the living room. Her face was bright red and she was obviously really annoyed by something. Jim had no idea what.

'Do you realise they've unpacked half the stuff that I had packed up to give to the charity shop? It's all over the floor. I told them they couldn't play with all that old stuff any more.'

'Well, I didn't know that, did I?' said Jim, shifting uncomfortably in his easy chair.

'You don't know about anything that is going on around here. None of it matters to you. It's just not your problem, is it?'

'Oh, Jan, that is so unfair. You know how busy I've been at work, it's...'

Her retort cut him off like a whiplash. 'I don't have time to spend arguing with you.' She stormed back upstairs to sort out the boys and Jim knew that the conflict between work and home had raised its head again.

Jim felt shell-shocked as he sat thinking after Jan's outburst. She was right, he should help more but when did he have time for anything after all the hours he put in for the company? Except for attending meetings with builders and bank managers, that is. Strange though, he mused, how the boys wanted to resurrect the games that Jan had decreed they would not be taking to the new house. A bit like the whiteboards and manual records that the staff at work were clinging to like life rafts after a shipwreck. With that thought he was lost again in deliberation of how he could deal with matters at work.

Later that evening Jim knew that he had some bridges to build. He made a quick call to Melanie, their regular babysitter and arranged for her to come round to look after the children. He then booked a table at Jan's favourite restaurant.

The sat down at the candlelit table and Jim announced 'Right, you have my undivided attention – unless I decide to have the Sticky Toffee Pudding which might distract me a little – and I want to hear all about the house and how I can help.'

Jan smiled indulgently at her husband 'I knew there was some reason why I married you.' She paused and Jim thought that maybe he was off the hook. 'But you needn't think that one meal will make up for all your neglect. We need to pull together on this move. I need more from you than a quick meal on a Saturday night. I need some real help, Jim. Mind you, my Mum and Dad are coming next week – in case you had forgotten.' She looked at Jim who did his best to hide his guilty look. She saw it but decided to ignore it. 'They will be a really big help to me. Mum's a dab hand with wrapping paper and Dad will be able to do most of the big jobs. It will certainly be a bit crowded in our house this next two weeks but I can't wait to see them again. It will be fun.'

'Mmm. Can't wait.' He mumbled.

'Anyway, let me tell you about the carpets we've ordered. Emma was so pleased with the ultra-modern one that she's picked out. She says it's 'cool' and she tried so hard to pretend she was used to doing things like choosing carpets but I could tell she was excited.'

They talked and talked and had a lovely meal. By the end of the evening, Jim had, he hoped, reassured his wife and had realised that he needed to divide his time between work and home a little more fairly. He knew that it was not going to get any easier for the next few weeks.

Monday morning saw Jim getting the details from Graham of the resolution of their latest crisis. They had had their first failure on the new system. A series of Purchase Orders generated on the new system and sent via EDI had failed. E-mails had been flying thick and fast between Abbey Martin's newly reorganised Purchasing Department and one of their longest-standing suppliers of raw materials. The orders had failed and the system had thrown up an exception report. No-one seemed to have any idea why the orders had been rejected by their suppliers and Graham had been tempted to tell everyone to forget the new system, just get on with placing an order in the good old-fashioned way so that supplies of the vital components could be maintained. Eventually, one of Abbey's most junior clerks had the idea to request a copy of the order received at the suppliers and – eureka! – the reason for the failure was obvious. The system had been configured to allow fourteen digits in the part number but any zeros at the beginning of the number were omitted on transmission. Result? – rejection by the supplier's system that deemed the first three zeros essential to understanding. Frank was called in to check the configurations and another problem was peacefully resolved.

As Jim was heaving a sigh of relief at their having solved this problem and Graham was preparing to go back to his own office, the consultant confirmed that the date for the next phase of the implementation had been decided upon. 'The 24th of next month, if that's OK with you Jim. Tony says we can go with that if we're all happy.'

'Sounds fine to me. Mind you, we'll probably be about ready to move in to our new house by then. I don't dare miss that or else my merger at home might fail!'

'I see. What do you want to do then?'

'Let's just see how it goes. We've already had two changes to the move-in date so who knows what will happen before then.' Jim had yet again put matters at home out of his mind and continued with his work. There was one thing he wanted to do today that he was really looking forward to. Graham's visit and his usual complaint about the lack of comfort in Jim's office had reminded him that he must order a new chair for his office. He turned to his computer and brought up the appropriate menu screen. With his system manual by his side he followed all the steps, logging in with his authority code and ordering a fairly basic chair. Within thirty minutes Jim had received an email confirming that the chair had been sourced from what he knew was their usual office furniture supplier and advising the expected delivery time. He knew that he would never even see the invoice for it as the system had been set up to send it electronically to the accounts department. Of course, his authority level meant that it would automatically be passed for payment. The new purchasing system worked like clockwork and Jim was impressed with the way everything linked in. He knew exactly when and how everything was happening. Having achieved success in that real-life test of the system, Jim delved deeper into the new purchasing system. There was still plenty of work to do in rationalising the company's supplier base.

By Friday Jim was on his way to report yet another exciting – if difficult – week in the implementation. He knocked on Tony's office door and, as usual, went in without waiting for a reply. He immediately regretted this. He was witness to a hastily abandoned, but obviously bitter, argument between Tony and John. John pushed past Jim and left the office without a word. An embarrassed Tony quickly recovered his composure and indicated a chair to Jim.

'Just a slight disagreement, Jim. Nothing we can't sort out.' he explained. 'Tell me how you've got on this week.' Tony smiled but Jim could see that the smile did not reach his eyes and there was a tired, resigned look on his face.

'I'll keep this brief, Tony. I can see that we've all had a hard week.'

'That's true. Tempers are starting to get a bit frayed around here, aren't they? But it's all in the name of progress so tell me what progress we're making.'

Jim purposely kept it brief. 'The budget is still OK, we've cleared up a few minor configuration problems and the only thing continuing to cause concern this week are the people problems that are beginning to surface.'

'People problems? What exactly do you mean? Not more staff with not enough to do, I hope.'

'No, no,' said Jim hastily, seeing the anger flare in Tony's eyes. Feelings were pretty close to the surface here, thought Jim. 'No, it's more the gossip that is going on. Lot's of muttering about the new system being 'useless' or 'too difficult' and how the old system was so much better. I suppose that it is all quite natural. Nobody likes change, do they? We just need to deal with that. We also need to sort out the way a lot of people are clinging to the relics of their old ways – the whiteboards, the manual logs, the spreadsheets that just check things that the new systems do automatically. While we allow that to carry on, we've got people wasting time.'

Tony thought for a few seconds. 'Yes, I suppose all those things are like security blankets, aren't they? Remind me, how are you planning to deal with all that?'

Jim summarised the programme of training workshops scheduled for the next two weeks. 'These should make sure that staff appreciate just what the system can do in practical terms and give them some reassurance. All aimed at working more efficiently. Then we can really get down to redeploying staff, rationalising our customer and supplier bases and then – you'll be pleased to hear – making some real savings. Of course, as you know we've not been waiting for the obvious benefits to come through from the new software. An interesting statistic that Graham threw at me the other day is that 80% of benefits from a re-engineering programme – which ERP is, of course – come from the non-system changes. Has he told you that one?'

'No, I don't think he's mentioned that specifically but I knew that ERP is about ways of doing things so it's no surprise that the benefits come, is it?'

'Well no, I suppose not but the magnitude certainly surprised me.'

Tony smiled and started to wrap up their meeting. 'That's great, Jim, thanks for that. Anything else?'

Jim hesitated but decided to plough on. 'There's one thing that I am at a loss as to how to deal with...'

'Fire away. I hope you know by now, Jim, that you can ask me anything.'

'There's a lot of gossip about Neville Crosland. I think because people don't know why he left so suddenly, they are making up their own versions of events.'

'I can certainly talk to you about that situation but do you really think that the staff need to know what went on?'

'Maybe not. They might not need to know exactly what went on but we need something to put all the speculation to rest.' He looked at Tony and was surprised to see uncertainty.

'In a nutshell, Neville left because he felt he couldn't cope with the changes. There was a lot more to it, of course, but that's it really.'

'I don't understand the need to keep that quiet – apart from being a bit discrete about Neville's shortcomings, that is.'

'Think about it, Jim. Neville was a director of the company and he was in charge of IT and Logistics for goodness sake! If he can't cope with ERP, how can we expect everyone else to get on with it?'

'Is that all there is to it?' asked Jim, not entirely convinced by Tony's explanation and getting a definite feeling that he was holding something back.

'I'll be honest with you, Jim, I felt really let down by Neville's departure. He just refused to take any part at all in the whole reorganisation of the company. I suppose I took it too personally, really.'

Jim could see the pain on Tony's face and hurried to bring this conversation to an end. He wished he had never started it. 'So what do you suggest we do about the gossip?'

'I really don't think that saying anything about Neville's shortcomings or my disagreements with him are going to help anyone in the company.' He was unmoved by Jim's doubtful look and continued 'These things die a natural death if they are just given time. My advice would be to let sleeping dogs lie. This is such an old issue. Why start worrying about it now? The gossip will die down, believe me.'

Jim realised that what he had just received from his boss was not advice but an order. He decided to leave it there. 'If you think it's best to do nothing, Tony, then that is exactly what I will do. I've plenty of other things to worry about. Let's just forget the situation with Neville.'

He discussed this conversation with Jan later that evening but he could tell that she was not really listening. She was more interested in setting their date to move in to their new house. Jim saw that she was becoming enthusiastic about the whole project again but felt compelled to keep her expectations within bounds. 'Let's just set a provisional date for now, shall we? We can make tentative plans then firm them up later when Ian gives us the go-ahead.

Jan looked crestfallen. 'But Ian has given us a date. How many times do you expect him to change it?' she said, her voice rising with indignation and disappointment.

He doggedly continued trying to make sure that he both managed her expectations and also kept his own options open. 'Being realistic, at this stage anything can happen. I know that from what's happened at work in the last few weeks. We need to be prepared for all sorts of changes. Let's just wait and see.' She still looked doubtful and he changed tack. 'You're doing really well with sorting out all our belongings, that will make our new life easier when we move.'

She got up and swept out of the room, calling over her shoulder as she went 'If you think that you can just change the subject, Jim Heswall, you're wrong. We're moving on the 24th and that's final.' She did not see the horrified look on his face. Jim picked up his newspaper, trying to put the problem out of his mind again. That's told me, he thought, I certainly hope she's not disappointed.

Points to reflect on

- Plan continuous training throughout the programme and after the programme is completed
- There will always be issues with data – even when things look OK
- Expectations need to be managed. Communicate often
- Don't let small things grow into major issues – sort them out with facts
- Remember that new processes take time to take hold and reach the efficiency envisioned
- Something new takes time to become routine. Every day people will become more familiar with the process and more comfortable with the system

Chapter 21 – Continuous improvement

'Would it matter too much if I was not here?' asked Jim. The other members of the Steering Group looked up in surprise and Jim continued. 'Our house move is actually scheduled for that date. I really must be there.'

Tony spoke up decisively while the others looked uncertain. 'Of course, you must be at home on that day. Don't worry about things here. I think you have done more than your fair share in this change project. Perhaps it's time that you concentrated a little more on the changes you've got at home.'

'That's true, Jim, and we have all done one implementation now so let's hope we have learned from our mistakes. This one should go like a dream.' added Graham.

The relief washed over Jim and he mumbled his thanks as the meeting moved on to plan the fine detail of the implementation of the two major modules.

That evening, Jim was pleased – and relieved – to be able to tell his wife and his in-laws – that he had managed to get the time off work to help with the move.

He looked around at his wife and her parents and all four children squeezed around the table. 'So, after all the work that you have all been doing, will there be anything left for me to do?'

'Oh, don't worry, we will save plenty of work for you,' answered his wife 'but you seem to have missed most of the clearing out of all the old junk from our cupboards.'

Her mother interrupted 'You'll never guess what we found today in the kitchen. I'm surprised she still had it.' They all waited for her to get to the point. 'That teapot that I bought for a joke when the girls were little. It was tucked away at the back of the crockery cupboard. What a monstrosity.' This was met by even more blank looks from Jan's father and from the four children but Jan and Jim looked uncomfortable. Jan's mother was laughing by this time and could hardly speak. 'You know the one, it was in the shape of Goldilocks and the Three Bears. I don't know what possessed me to buy it – I suppose I thought the girls would like it...' She tailed off as she could see that the children didn't know what was going on. Jim and Jan had started to laugh with relief along with her. Even so, they did not dare to agree that the teapot was indeed a monstrosity. Jim struggled to keep control. If only it was so easy to persuade people at work that they didn't need the relics of their old ways.

The latest implementations – the move at home and the two major modules at work – went much more smoothly than expected. A post-launch meeting was planned so that the Steering Group could start to realise some of the benefits of the changes while a housewarming party was planned at the Heswall's new home.

Graham took the lead in the 'Continuous Improvement' meeting at Abbey Martin. 'We need to get the rationalisation processes moving along so let's go through the different processes and cut out some excess capacity.'

Jim took over. 'Let's tackle suppliers first. I've rationalised our supplier base and I think it's in pretty good shape now. There will no doubt be a few things we can tweak here and there but there are only about half the number of suppliers now compared with the combined lists from the two companies.'

They went through the new supplier database, the authority levels that had been set and the savings that had already been made on the new prices that had been negotiated. Graham then moved on to the Order Management system.

'For such a large implementation this has gone really well. We must have learned our lessons on the first two modules.' He smiled at everyone around the table.

'Seriously, well done everyone. We still have a lot to do but so far, so good.'

Tony spoke up 'Graham's right. I know that this is the result of a lot of hard work by everyone here. Thank you everyone. Sorry to interrupt, Graham. Carry on.'

'Thanks Tony. Let's review the items that are still outstanding on the two modules then we can plan the customer rationalisation. That's where the major savings will come from these modules.' He glanced at John – Abbey's Finance Director and at Bill, Martin's Accountant 'The other aspect we need to look at today is the quality of the financial detail we're getting from the new modules. I think you'll be impressed with these.'

They quickly reviewed the items that Graham had mentioned then allocated specific tasks to ensure that nothing was missed in the search for savings.

At the end of the meeting Tony stood up and cleared his throat. Jim could feel that there was a major announcement coming and he looked around everyone else. With the exception of John, all eyes were on Tony as they waited, anticipating the news. As Jim studied John's face he suddenly realised that the news was serious and involved this man who he had worked closely with over the last few months and who had become his friend. Jim dragged his thoughts back to Tony. 'And so, it is with regret that I have to announce the resignation from the board of John Forsythe. I am personally sorry to lose him and know that the company will miss his expertise and calm management. In the short term we will all have to pick up the extra duties involved and further announcements will be made within the next few weeks. I will be joining the groups and committees for the business change projects alongside John until he leaves – just to ensure some continuity. ' He sat down heavily and studied the papers in front of him. Jim watched him and was not certain whether it was embarrassment or pain that kept him from saying anything more. He glanced quickly at John but he was keeping tight-lipped.

The meeting broke up and both Tony and John quickly went their separate ways so Jim did not get the opportunity to find out more. From their demeanour, he was doubtful as to whether either would say more on the subject of John's departure.

Later that day, he got the chance to bring Jan up to date with the events at Abbey Martin. 'I haven't been able to speak to either of them so I don't know what to think.'

'I hope this doesn't mean even more work for you.' said Jan, with her usual concern for her family's welfare.

'No, I don't think it will but who knows? Anyway, John is carrying on for a handover period of three months so the ERP and most of the major changes that go with that will have been sorted out by then. I'm hoping things will get easier from now on.'

'It's about time. I know you managed to get time off work to do the actual move but be honest, you haven't given it your full attention, have you? Not really.'

He took a breath, preparing to mount a robust defence but Jan continued 'No, don't get defensive, I'm not complaining. I'm proud of you. I think you've done a great job from what I've heard. It's been a very steep learning curve, hasn't it?'

Jim was relieved to be back on the safer ground of changes at Abbey Martin but knew that Jan would notice – and not hesitate to let him know that she had noticed – if he was too eager to leave behind any discussion of the move. 'Talking about the move, now that we're fairly organised here don't you think we should get on with planning the housewarming party?'

Her face lit up. 'I'm looking forward to it. Who should we invite?'

'Well John and Brenda, obviously. I did think that maybe Tony and his wife would come along too but

I'm not so sure now. I'll have to wait until I've found out more about what went on between them.'

'Typical! That is just so typical of you recently. You can turn any discussion into an analysis of Abbey Martin's problems.' As Jan got up and flounced out of the room, Jim reflected that, yet again, he had said the wrong thing. He followed her to the kitchen.

'Let's not argue, Jan.' he said as he put his arms around her 'We're so lucky – look at this dream kitchen we've ended up with.

She flicked drops of water at him from her soapy, wet hands. 'Not much thanks due to you there.'

'I'm just doing my best. I think I'm better at change at work than change at home.'

She shrugged her shoulders. 'Men!' was all she said.

Three weeks later the party was organised. John and Brenda had been invited and Jan's parents were already scheduling their first visit to the new house to coincide with the party. Despite Jim's best efforts the reason for John's imminent departure was still not clear so he had not invited Tony and his wife. Tony had kept to his policy of not gossiping about fellow directors and John had merely said that 'his job was done and it was time to move on.' Discussions with Graham and Frank had not revealed any definite details about the situation between John and Tony but they both agreed that the departure of key players in the change process was a common feature of this type of project.

With the major modules of Order Management and Finance up and running, there was a lot to do. Jim was not on the Order Management team but with John's imminent departure, he was seconded on to the Finance Team and became heavily involved with work in that area alongside Bill from the old Martin accounts department. One meeting saw the two teams come together for customer rationalisation. With Graham leading the meeting and Sales Managers from both companies involved, they started to tackle the method for deciding which customers should be gently jettisoned and which should be dealt with in different ways so that the company could continue to supply them but without the associated high costs.

Graham started the meeting exactly on time. 'Firstly, let me restate just why we need to reduce this enormous list of customers.' He glanced meaningfully at the silent sales staff and indicated the thick computer print out of Abbey Martin customers. This had just been produced from the new system and included an amalgamated list of customers form the two companies together with ratings based on their usefulness to the company. This issue of usefulness had been a hard fought area in previous meetings. It was obvious that the sales staff did not want to lose any of their customers and were particularly attached to their old favourites. Some of them even seemed to have difficulties appreciating that profitability had to be at the heart of any discussions of change. Graham continued 'The analysis that we have been able to get from the new system shows us that at least fifteen percent of the customers on the list cost more to service per year than the company could make over five years – even if they were to increase their average order value substantially. They must be taken off the company's customer base. Keeping them on as they were just cannot be justified even if you calculate their value to the company over a long period of time.' He looked up as Abbey's sales manager took a deep breath and prepared to launch a defence. Graham quickly carried on. 'There can be no exceptions to this. Abbey Martin is a company that will not – I repeat, will not - continue to deal with companies from which we cannot make a profit.' The sales manager clamped his lips tightly shut but Jim could see that he had not accepted the issue and attempted to move the meeting on.

'Can I just add something from the point of view of the Finance Team?'

Graham smiled at him and signalled for Jim to continue. 'We must make sure that we don't cut these customers off – however, we decide to do that – without first making sure that there are no old debts on the system.'

'That's a good point Jim. Thank you.' Bill added gratefully 'The much improved information on the system now gives us all we need to know to get these old invoices up to date but my credit control staff are still having a hard time making some of these customers pay up.' He studied the list of customers and was about to start on specific examples when Graham interrupted.

'Of course, debtor days was one of the factors that we used when we calculated each customer's usefulness. Many of these customers seem to just use us as a source of free credit. Can I suggest, Bill, that you run a report on the customers that we've selected to get rid of. Include all the details of what is outstanding and give it to the sales people to chase up the payments for you. You can report back on progress at the next meeting.' He then nodded to Jim 'Rest assured, we won't do anything final until we've resolved that problem.'

The sales manager could contain himself no longer. 'We're turning into debt collectors now are we?'

Graham shot back 'Perhaps we're just turning into a profitable company.'

John intervened at this point – up to now he had been observing quietly. 'Let's take a break there, shall we? We've still got plenty to get through so can we have everyone back in here after coffee – say twenty minutes.' He got up and walked out.

When they resumed, the atmosphere was still strained and they got through the next part of a very uncomfortable meeting by lunchtime. By then they had managed to rationalise the customer base with firm plans to jettison most of the unprofitable customers. The sales people had recovered their composure. They had agreed, and added to, the arrangements that had been put in place to deal with the smaller, but viable customers via a newly created distributorship run by one of their major customers.

Graham wound up the morning's efforts 'Thanks everyone. I know this type of action is never easy. I think we've done a good job there. Let's get some lunch and be back here at two.'

Jim managed to collar John as they left the meeting room. 'Fancy a quick lunch, John?'

John appeared to be considering refusing but turned to Jim with a smile. 'Thanks, Jim. That would be great. Just a sandwich, eh?'

They walked briskly to the pub, making small talk about the weather and their families until they sat down at a corner table, having obtained beer and sandwiches.

Jim took a large bite of his tuna mayonnaise and wondered how he could tactfully broach the subject of John's resignation. He was saved the effort as John said 'I suppose you're wondering what's happened between me and Tony.'

Jim nodded, his mouth full.

'The short answer is nothing. I'm going into semi-retirement.' seeing Jim's look of scepticism, he continued 'I think I mentioned to you that Tony had headhunted me for this job?' Jim nodded. 'I was about to change direction then but I agreed to do this job as a favour to Tony. We've been friends for a long time.'

'So you've not had a bust up?'

'No, not really.' Smiled John 'He isn't pleased about it. Thinks the job isn't over here but he knew that I only accepted the position for a very limited time – just until the merged company was up and running so he'll get used to the idea. You'll find that he can be a bit like a spoilt child when he doesn't get his own way but he'll get over it. He will move on to his next challenge soon. That's how he is.'

'I see.' said Jim. 'So what are you going to do?'

'I'm tying up the loose ends at Abbey – sorry, Abbey Martin – then I'm going to do some consultancy work. Freelance, you know. There are one or two people who I've worked with in the past who want me to head up projects for them. Even Tony has mentioned a possibility in the future. I'm looking forward to it.'

They went back in to the meeting, knowing that the work for the afternoon was a lot less demanding than the morning's agenda had been. They ploughed through figures on query handling by the customer service team and details of changes made or still to be made to authority levels. Graham again summed things up at the end of the meeting 'I hope we can all agree that there has been significant improvement already in our query handling and...' He stopped, as he spotted Diane, the Customer Service Manager, anxious to make a point.

The normally serious-looking woman in a navy suit and severe white shirt smiled as she said 'I just wanted to say, Graham, that I know how difficult my staff have found this whole process but we can already see the benefits. The information we possess now to deal with even routine enquiries about orders mean that we can do a much better job – and do it quicker. My staff have commented that they are feeling more professional. In fact, one girl said a couple of days ago that with the authority she now has to deal with situations and the fact that customers knew exactly who to contact, she is doing a better job. The feeling has been for a long time that 'the customers must think we're stupid'. That is very demoralising and I'm glad we've sorted that out. I can see this whole thing making my job easier and helping us to keep the customers.'

'Well, what more can I say? That is exactly what we've been aiming at. I appreciate that, Diane, thank you. I realise that not everyone is as happy as you and your staff are with the new ways of working but all the changes are aimed at making the company more profitable and ensuring our survival. If anyone doubts that, they should speak to me or to Tony.' He looked sharply at the sales staff ranged along one side of the table. They all looked down at their notepads, saying nothing. 'Does anyone have any more to add before I close the meeting?' He was met by silence apart from the sound of papers being gently shuffled but no one commented further. 'OK thanks everyone.'

Jim heaved a sigh of relief as he got wearily into his car straight after the meeting. When he walked through his front door he was met by the sound of bickering and slamming doors. He briefly considered turning around and going for a relaxing drive but as he was hungry, he stayed. He found Jan upstairs unpacking boxes of clothes in their bedroom. 'What's for dinner?' he asked.

This remark was answered with a barrage of socks as Jan threw at him the contents of the box she was emptying. 'That is just what I need. You asking silly questions.'

'Had a hard day then?' he asked.

She did not answer that question either but continued, grim faced, to sort out their belongings.

'Another daft question, uh?' He sat down on the bed.

'Don't sit down. Make yourself useful.' His wife shrieked.

'Jan I am just too tired for this.'

'And you think I'm not? Andrew has done nothing but complain since he got home from school, Emma is sulking because I won't let her bake a cake and the other two are bickering about who's turn it is to wash up after dinner. That's if we ever get any dinner. At this rate it will be time for bed before I've made any progress at all on these boxes.' She paused for breath and Jim leapt in.

'I'm hungry so I'm going to get us a meal. What shall I make?'

'Why is that decision my responsibility?' his wife snapped, her voice muffled as she poked around in the bottom of yet another box.
'Only asking. I'll go and check.' He went back down the stairs, shaking his head and thinking he just could not win.
'There's salad in the fridge.' came a voice from the bedroom.
He stepped into the kitchen and stopped. 'Hey, the new fridge has arrived. Wow, that looks impressive. Much better than the old one.' He nudged Andrew who was sitting at the breakfast bar eating an apple. 'Have you tried out the new icemaker on this thing?' asked Jim as he fiddled with the switches on the front of the new, giant sized fridge.
'No, Mum says we're not to mess with it.' answered Andrew, getting up eagerly and standing beside his father.
Jim immediately put his hands in his pockets. Jan was in a bad enough mood already, he thought, I'd better not be caught giving the kids a bad example.
'Right then, you and I are in charge of dinner. We're having salad apparently.'
Andrew screwed his face up and Jim echoed the thought but continued removing the relevant items from the fridge. 'Get some plates out, please' he instructed. The boy stood on tiptoe as he opened all the cupboards alongside the kitchen, slamming each one shut as he failed to find what he was looking for.
'Can't find any, Dad.' he announced, sitting back down at the breakfast bar and watching as Jim opened a few wall cupboards and found some large plates.
'It's rubbish this kitchen, Dad.' the young boy declared 'Our old one was much better. I could find anything in that kitchen. Plates, cups, glasses, knives, forks…'
'That's enough Andrew.' said Jim sharply 'We just need to get used to this one, that's all. It's a lovely kitchen. Emma and your mum like it.'

Andrew said no more but the look on his face betrayed his opinion of the changes. Jim sighed. One day he would think about exactly why everyone was resistant to change but for now he just wanted to get a meal organised and sit down for a rest.

> **Points to reflect on**
>
> - Remember to budget to invest in more training – even repeated training
> - Changes will continue after implementation. Benefits will continue to build
> - Be prepared for more people issues – nerves will fray
> - Communicate the long hours that will be needed, but that this will begin to diminish
> - Ask people using new processes for new benefits they are developing and publicise these benefits to encourage best practice
> - Develop a recognition and rewards programme. Budget for this at the outset

Chapter 22 – Lessons learned

The most difficult period of three months in the business, which Graham had predicted right at the start, was nearly at an end and life was getting a little easier at Abbey Martin. People were getting used to their new ways of working and many of the wrinkles in the new system had been ironed out. Nevertheless, Jim knew that the changes were not over. He was right. At their routine Friday afternoon update meeting, Tony went through the figures about cost savings with more than his usual attention to detail. Jim knew that further personnel rationalisation was imminent and that many people would either lose their jobs or have them radically altered to ensure that the cost savings of the project could be realised. He decided to mention his worries to Tony.

'I know that changes are inevitable, Tony, but I can't help but worry about them. Job security is important to me.'

'You? Oh, you'll be OK. Don't worry.' he replied, dismissively.

Jim looked at Tony uncertainly but could see that further discussion of his position would not be welcomed at this stage. Tony's mind was obviously on something else. He did not have to wait long to find out what it was.

'I've set a date for the final move to our main premises. The Martin manufacturing and warehouse facilities will close in six weeks. The office staff will follow the week after. As you know the new warehouse at Abbey is nearly finished. We might have to lease a bit of extra space for a few months until we've worked out the redundant stock situation. I've still got hopes of selling most of that at a good price. And, oh yes, I wanted to let you know that Ron is considering early retirement. We've made him an offer.'

'I see. It's been obvious from the beginning that he could not cope with the new systems. I think he's allergic to computers or something.' Jim laughed to lighten the moment.

'Yes, I suspect you're right. Whatever, there's no room at Abbey for people who refuse to change.' Tony said, his voice unusually cold and hard. 'You're not frightened of change, Jim, are you? That's become obvious too.'

Jim hastily agreed.

'We'll talk about how all this affects you at a later date.' He suddenly seemed to notice Jim's discomfort and worry and was quick to reassure. 'Don't worry, Jim. I'm just a bit preoccupied at the moment but things will work out, you'll see. For now though, it is other people that I want to mention to you. The move will bring personnel changes to the fore.'

'Yes, I was rather afraid of that.' said Jim, thoughtfully, waiting for Tony to elaborate.

'The integration of the two warehousing operations is just the tip of the iceberg, Jim. There will be redundancies there, of course. I've also got lists together of staff in Accounts, Human Resources, Customer Services and, of course, the sales staff.'

'I knew this was coming but it doesn't make it any easier when it finally comes does it?'

'It's never easy. Just remember that the potential savings we could make were the whole point of the ERP project. We just need to have patience when it comes to the results. I must confess I can be the world's worst when it comes to patience. I want to get things done, see the results then move on to the next project.' He became brisk again. 'Now, what are the figures like on the Customer Service queries this week? Are we getting to grips with the changes?'

They went through the details in record time. Things were beginning to run more smoothly, although the unrest in all departments was causing a number of personnel issues. As he reported to Tony, Caroline from HR was working flat out to resolve these issues and to help all the managers get their staff motivated.

She had put a lot of effort into making sure that all the departmental managers knew the upside as well as the downside of all the changes. Whenever she heard a positive comment from someone who perhaps had opted for voluntary redundancy and was relieved to be 'getting out of the rat race' or someone who was thrilled with their redeployment and enhanced job prospects, she was quick to spread the word. Caroline's propensity for gossip was finally paying off for the company.

'Oh, I know how hard she's working. We've all had to put a lot of work into this. You've certainly had to do a lot of things that weren't familiar to you. You've no regrets, have you?'

'Regrets?' said Jim, in surprise 'Certainly not! I've found it hard going – it has been a very steep learning curve for me but I've enjoyed it. Well, mostly.'

'How is the house coming along? I know you've had your challenges at home as well as at work over the last few months.'

'I do feel a bit guilty about not helping my wife as much as perhaps I should have but she seems to have coped. Things are going well. In fact, we're having the housewarming party tomorrow.' He had been meaning to invite Tony since his conversation with John had reassured him that there were no hard feelings to spoil the atmosphere and now he grasped his chance. 'Why don't you and your wife call in to help us celebrate tomorrow evening? John and Brenda Forsythe will be there, I think Graham will be able to make it and plenty of others.'

'That will be good. I will check with the boss.' On seeing Jim's quizzical look, he explained 'That's Louise – my wife.'

Yet again when Jim walked into the house after work, the house was in chaos. Jan's parents had just arrived, the boys were in the garden getting muddy while Jan was shouting for them to come in and get washed to greet their grandparents and the girls were in the kitchen with packages of food all around them.

Jim said hello to his in-laws 'Welcome to the new house. I'm certainly glad to see you. We need all the help we can get to be ready in time for tomorrow's party.'

Jan's mother rolled up her sleeves saying 'Just tell me where to start. Bob...' she called to her husband 'Bob, take our bags upstairs and then we'll sort out a meal for all of us. You don't want to be cooking tonight.' She eyed the kitchen where the girls were trying to squeeze all the shopping into the numerous cupboards. 'I'll set to cleaning the bathrooms in the meantime, if that's all right with you dear?' She glanced sympathetically at her frazzled daughter as she came back into the house. Jan nodded.

Jim could see that it was his turn to report for duty 'I'll just have a quick shower then I'm all yours, Jan. Just tell me what you want me to do.'

Despite the chaos of Friday evening and the few weeks since they had moved into the house, Saturday, although hard work, felt more organised and Jan was noticeably happier. Everything was ready on time and the guests started to arrive shortly after. The children had been allowed to take part in the first couple of hours of the party on condition that they helped with the drinks and the food. It was obvious that they were enjoying themselves and Jan felt herself relax and start to take an interest in getting to know some of their new neighbours and also some of Jim's work colleagues.

Jim drew her across the room to meet Tony and Louise. 'Pleased to meet you.' Jan smiled at the couple then turned to Louise 'Has your husband been as preoccupied as mine?' she asked.

'He's always busy but then, we haven't moved house during the merger so you've got my sympathy. It must have been tough on you. It's a beautiful house, by the way. You must have worked very hard.'

Jan grinned. 'You don't know the half of it. It's been a nightmare.' Meanwhile Jim and Tony were standing to one side discussing Jim's plans for the garden.

'You won't be doing it all yourself, will you Jim? Looks like a lot of work to me.'

'No, Jan's the creative genius in this family. I think there are some very good landscape gardeners locally so they will be doing the heavy stuff.'

'I'm glad to hear that. I was starting to feel guilty. Louise runs the show at home. We've got plenty to keep us busy at Abbey Martin. Isn't that true Jim?' He nudged Jim and grinned at Jan and Louise. Both women shot looks that would kill but Jim was pleased to hear his Managing Director reaffirm his involvement in the company's future plans.

Jan saw a few people standing around without a drink. 'Would you excuse me a second please, Louise.' She walked rapidly to the kitchen where Andrew and Philippa were in conference beside the fridge. 'What's happening? There are guests without drinks out there.'

Both children started to speak at once. 'There's no ice.' said Andrew.

'This silly new fridge won't work, Mum.' shouted Philippa 'It's useless. At least our old fridge could make ice cubes.'

Jan had a quick look at the complicated looking apparatus on the front of the new fridge and realised that they weren't going to get any ice out of that tonight. She then remembered that they had thrown away the old ice cube trays when the new fridge had been delivered. Her father wandered into the kitchen and was immediately despatched to buy a large bag of ice cubes from the convenience store.

The doorbell rang just as Jan was going past. It was John and Brenda. 'Lovely to see you again. Go through' she said directing them to the living room

'I'll be in just as soon as I've sorted out our latest crisis. We can have a good chat later.'
John and his wife entered the living room and were soon deep in conversation with Tony and Louise. They were obviously old friends. Jim had a brief word with them all but could see they were getting on well so he carried on circulating and getting to know his neighbours. The evening progressed and the children were despatched to bed by their grandparents and most guests, including Tony and Louise, made their discreet exits.
It was very late before Jan and Jim managed to sit down with just a few friends, including Graham, John and Brenda, and all admired the new house.
'It can't have been easy, though, Jan.' said Brenda.
'You warned me it would be painful but I'm not sure that either Jim or I really believed you.'
Jim agreed. 'I certainly had no idea how many things could go wrong or how difficult it would all be. We did eventually get someone competent to organise the site for us. Ian has been a godsend, John. Thanks for recommending him. Mind you, it wasn't all plain sailing even after we had got him on board. We had delays and budget increases. I'm just so glad that it's all over.'
Jan laughed. 'Over? That's what you think. The new fridge doesn't work. It won't make ice cubes. The kids nearly staged a walkout earlier! We'll have to buy some new ice cube trays. We've thrown the old ones away.'
'No way!' exclaimed Jim 'We absolutely must make the new method work. I'll ring the suppliers on Monday. The ice cube tray system is a re-engineered process so we must use the new way of working. We wouldn't give in so easily with a new system at work, would we John?'
They all laughed and Jan looked dumbfounded 'Only you could turn a problem with our fridge into a lecture

on your ERP project! That just about beats everything!'

The conversation soon moved on to a more general discussion of the situation at Abbey Martin. Graham said cheerfully 'It will soon be all change again at Abbey Martin. You're moving on in a couple of months aren't you John, and Frank's period of post-implementation work ends in three weeks. My contract comes to an end soon but we'll be back for the other modules – when Tony gives the go-ahead on more expenditure. I suppose he will have to see the benefits of this project first.'

John was first to respond 'There will be a lot more departures before the next modules get implemented.' He noticed Jim and Jan's serious faces and smiled at them reassuringly 'Don't worry. I think your position will be OK, Jim. You're moving to your new office soon aren't you?'

'Yes, I'm almost there already. I've been working more and more at the main premises lately. It's a much nicer atmosphere there so I won't be too sorry to leave the Martin premises. I've got a temporary office in the main building but I'm not sure where I will be located eventually. I just hope they find me a permanent home.'

'I'm sure they will.' commented John 'Maybe you'll get my old office.'

'That's a lot more luxurious than I've been used to – either as a Purchasing Manager or a Projects Manager – so that would be nice.'

Jan chipped in 'You deserve a secure job and a nice office, Jim. You've worked really hard.'

John nodded in agreement and Jim basked, momentarily, in the glory.

'It's a bit like the house move though, it has been painful too. I'm not sure you warned me about that, John.'

'Would you have taken any more notice of him at work than you did at home?' asked Jan.

Jim laughed. 'Probably not. Somehow, enthusiasm for change takes over doesn't it? Then you forget – or refuse to see – that some of it will probably be painful.' His face brightened. 'I know that things are already starting to feel more positive. The whole thing is based on new processes and values and we've got a lot to look forward to. Sure, some of the redundancy situations have been difficult but I've been surprised, I must admit, that quite a few people are viewing redundancy as a positive thing. Well, like you for instance – you're happy to finish the job and move on aren't you?'

'Yes, I am. You're right, Jim,' answered John 'and Tony has a reputation for holding on to the best people. Besides, Graham has already done his assessment centres to ensure the right blend of skills are retained. I think all that was pretty much sorted out at the design stage.'

Points to reflect on

- There will be some painful changes that are for the general good
- New structure, new people and new policies will encourage new management styles
- New management styles should drive continuous improvement
- Use new systems to facilitate continuous improvement
- Assess the results. Review the achievements versus the plan
- Getting the right people on the project will drive more benefit than you originally envisaged
- Be prepared to manage the negative people from Day 1
- Identify the key influencers and ensure that they are 'on board' from Day 1
- The system and process will never be perfect. That is why you need continuous improvement

Chapter 23 – Realising the Benefits

The Sunday morning after the housewarming party was spent with everyone pitching in to help tidy the house and culminated in Jan's parents taking them all out for lunch.

Jan's father, Bob, proposed a toast 'Madge and I just want to say how very proud we are of all of you. I must admit that I was a bit doubtful when you started. Not many people build a house from scratch and I thought that you had bitten off a bit more than you could chew.'

'You and me both, Dad.' agreed Jan.

'But you've done it. So well done the Heswalls!'

'Well done the Heswalls' echoed the brightly beaming children.

'Thanks for your help. We've got a lovely home here now – one that will work perfectly for our family.' said Jan as she looked around at her extended family.

'We haven't finished yet, have we Dad?' said Andrew, barely able to contain his excitement.

'Well, I'd like to think that we can have a few days rest, Andrew.'

'No, Dad, I mean the swimming pool.'

'I know perfectly well what you mean. We've got it in the plans – it's down under continuous improvement.'

Andrew subsided, satisfied that his Dad was with him on the pool development.

'Does this mean that you won't be late for work any more, Dad?' asked James

'I can't promise that I will never be late ever again, James but at least I won't be able to blame anyone else for being in the bathroom.'

'You've got three to choose from now.' said Emma.

'I think I'll stick to the en-suite by our bedroom.'

'Yes, that's my favourite one. The blue colour scheme is so calming.' said Madge. She saw the look on Emma's face and quickly added 'Your lemon and lilac scheme is beautiful too. It's all lovely. Anyway, this house is just what you all deserve.'

'It will certainly make life a bit easier.' said Jan.

'Did Mum tell you that it's all environmentally sound?' asked Emma

'Is it really? How did you manage that, then? Was that your idea, Emma?'

Emma agreed, shyly, that it was. 'We've studied a lot about ecologically sound investment and projects at school. Its...' Jan could see her mother's eyes beginning to glaze over and was relieved when Emma's lecture was interrupted by the waiter's arrival with the food.

Madge turned to her daughter and whispered 'I'm sure all this environmental stuff is marvellous, dear, but I'm more concerned about you and your marriage. That's where this house could make the difference really. You were a little cramped in your house, weren't you? But I can see you're happy and I'm so glad for you.' Jan knew her mother had a good point and remembered again what all the effort had been for. It was great to have a fancy house but the real benefit was the removal of the hassle caused by living in a house that was too small. As she had thought so many times, this new family was too precious to put at risk and all the effort was already proving worthwhile.

They carried on long into the afternoon, discussing the finer points of the house and how it would make their relatively new family life work so much better.

The following day, Jim was able to set off for work at a leisurely pace. He had no problems with the bathroom these days and knew that Jan was happy with the way things had turned out. He also thought that he had a relatively easy day at work ahead of him. There was a presentation scheduled for the afternoon in which Tony and Graham were to give details of the benefits already realised from the implementation.

When he arrived at work he got busy reorganising his office so that he was a bit nearer being ready for the big move. While he was doing that he wished, not for the first time, that he knew what sort of job he would have in the future with Abbey Martin. Little did he know that the news he wanted – and some that he did not – would be delivered to him within the next few hours.

His telephone rang and when he answered it he received a curt instruction from Tony's secretary to see Tony in his office at noon. During the morning, he worked in various locations within the factory. First he was in Purchasing sorting out what turned out to be a minor problem, involving someone not understanding a part of the new process. Then he was in the Manufacturing Department collecting the latest data on shop floor accident figures in preparation for the Realisation of Benefits meeting that afternoon. He finished the morning in the Customer Service Department seeing how the staff were working with the new system. In every department he discovered an uneasy atmosphere filled with gossip. He was eventually forced to listen to the talk on the shop floor when David Hardcastle, the Manufacturing Manager, asked him whether he still had a job.

'What's up, David?' he asked, when he heard the aggressive tone in the Production Manager's voice.

'It's not easy, you know, dealing with redundancies.'

'Don't you think I know that?'

'You appear not to. A lot of people are blaming you for this, you know. I happen to think that's a bit unfair but I'm just passing on what I'm hearing.' Jim was too rushed and also too nervous, knowing that his fate was likely to be decided very shortly, to respond right away but resolved to get back to this situation as soon as he could. Continuous improvement, he reminded himself as he left the shop floor.

As noon arrived, he entered Tony's office somewhat nervously. Here goes, he thought.

'Oh, hello, Jim. Come in and sit down. Let's have a chat.'

Ah, not a good start, thought Jim. He was seeing warnings in every word now. 'I need to talk to you about your position in the company.' Jim's heart sank even further. 'I have been very impressed with how you have handled everything over the last few months. I know it has not been easy. Now, Jim, after all that has gone on recently, do you see your long term future with Abbey Martin?'

'I would like to think so.' he answered carefully.

'Good. I want to see you here for the foreseeable future. As you know, we've had a lot of personnel changes. John is taking semi retirement and several more changes and redundancies have just been announced. Are you up to speed with all those?' Without waiting for anything more than a slight nod from Jim, he continued 'Two other people who you have been working closely with will shortly be going on to other things – Graham and Frank. So what have we in store for you now?'

'I just hope I'm going to like what you've got to tell me. The suspense is killing me.' He finally admitted to Tony with a grin.

'Ok Jim. What I would like to offer you is a directorship.' He paused and Jim stared at him open mouthed.

'I would like you to take on part of Neville's duties. I want you to be IT and Logistics Director. How does that sound, Jim?'

'That sounds, erm... that sounds just wonderful.' Jim could hardly think, he was so taken aback.

'We can discuss terms of course but rest assured it will be a considerably better package than you have been used to.' Jim stayed silent, not trusting himself to say anything sensible. 'And of course, you will need to sort out a better car and so on. I've earmarked an office at Abbey for you – John Forsythe's old office, just down the corridor from mine. Do you think you will be able to accept – in principle?' he asked with a smile. He got up and held out his hand to Jim 'I'm taking that for a

yes, Jim. Welcome to the board of Abbey Martin. Do you fancy some lunch?'
'That would be great. I'm not sure that I could eat right now – my stomach is doing cartwheels -and of course, I'd like to ring my wife first but yes, lunch would be fine.'
'Right, I'll see you in Reception in twenty minutes?'
Jim got up and left Tony's office as if sleepwalking. It was only when he managed to get hold of Jan on the phone at work and had put the news of his appointment into words that he realised how much of a step forward he had taken.' Jan was screaming, crying and laughing all at the same time and it was some moments before Jim could get any sense out of her. 'So I'm off for lunch with the Managing Director now,' he said 'we can talk tonight. I'm glad you're pleased.' He put the phone down, took a deep breath and went for lunch.
At the meeting in the afternoon, Jim was surprised to hear Tony start the meeting with a special announcement of the new director's appointment. The congratulations from his colleagues made it all seem more real. Jim Heswall, IT and Logistics Director. He rolled it around his tongue. Tony also summed up the latest redundancy situation and an altogether more sober atmosphere descended on the meeting. Undaunted, Tony continued.
'We must all keep in mind that the changes to personnel are a part of what we started all this for. We all knew that the road would not be easy and in some cases, unfortunately, it has been extremely painful. However, we have been successful in accomplishing the implementation and we have realised the savings we expected at this point. There will be more to come but for now, Graham is going to give us some more information about the deliverables of this project. Graham, over to you.' Tony sat down.
'I can only follow that, Tony, with some figures about the profits we're looking at in the next few months.'

He proceeded to give his usual presentation involving charts and graphs. They showed more profits from the same sales and also a projection of extra profits that would follow from the up-selling opportunities that were now starting to arise in the Customer Service Department.

'We've got a more motivated workforce in there now. They are starting to settle in to their new work routines and we are working with them to identify more opportunities for them to add extra sales to their routine contact with customers.'

'I believe that's working very well, Graham.' commented Tony 'Any other changes to report in that area?'

'Yes. The automated system for setting up new customers' accounts has been very well received. There's good feedback from new customers and even the sales staff are happy with the results they are achieving with it.' Several people around the table smiled at this point – the sales people were well known within the company for being hard to impress. 'The data that they are capturing right at the start of the customer relationship will no doubt prove useful in the future. The cost reductions there will start to show themselves over the next few months.'

'That's great, Graham. The important thing is that these systems have been successfully put in place so that we will realise the savings we are looking for.' said Tony. 'I believe we've already had some positive feedback on the automated e-mails we're sending that welcome new customers?'

'That's right. One customer in particular seems to think that this is revolutionary. He commented that it was a pleasure to spend his money with Abbey Martin!'

'That's a first.' muttered the sales manager.

'We have cut the number of staff in the combined Customer Service Department by a considerable percentage but I'm confident that they will be able to handle all the business that the sales force can bring in for the foreseeable future – and handle it to a much higher standard of service. For those of you who haven't been closely involved in the implementation – the improvements include automated welcome emails as Tony mentioned, the whole process of opening a new account is now computerised and it links in to all the other parts of the system such as accounts, sales, credit control, order processing and pricing and, of course, the customer is able to track delivery details of their order via their web portal. It will also link in to the Manufacturing Module when that is implemented.'

'Yes,' said Tony 'I think we can safely say 'when' rather than 'if' the next module is implemented. We will discuss the timing later, Graham.' he said, decisively.

Graham grinned broadly. 'I'm glad you're happy enough already with this implementation, Tony, to contemplate more software modules. At this stage some companies are so traumatised by the implementation process that they say 'never again' – despite the huge savings. I must say this has been a good implementation.'

'That is due not in small part to you and your staff, Graham, and, of course, we've got some hotshots here too.' He smiled around the table. 'Well done everybody.'

When he arrived home that evening. The house was exceptionally quiet.

'Are they all out?' Jim asked Jan as she walked into the hall to greet him.

'No, they're all in their rooms. At least we can get a bit of peace and quiet in this house. Good isn't it? How are you feeling now? Have you got used to the idea of being a company director yet?'

'No, not at all. It just doesn't seem real. Have you told the kids?'

'It's been difficult to keep it to myself but no, I haven't. I thought you would like to do that.' she answered,

hugging her husband. She moved to the bottom of the stairs and shouted for the children to come down for dinner. 'It's a special meal. We're celebrating so hurry up – and wash your hands first.'
The two girls were intrigued and came down the stairs within seconds. The two boys were, as always, slower to get to the dining table.
'What are we celebrating, Mum?' asked Philippa as she sat down. 'Ooh, this is nice. Candles and flowers! It feels like someone's birthday.'
'No, it's better than that.' laughed Jan
'Do we have to wait for the boys?' asked Emma 'They're hopeless.'
'This is important, Emma. Of course we have to wait for the boys.' warned Jan.
The boys came tumbling into the dining room. James excitedly reporting on the computer game he had been playing on the new computer that he no longer had to share with his brother.
As soon as they were all seated Jim made his announcement. It was met by blank looks from the boys but the obvious delight from Jim, Jan and the girls soon communicated itself and there were squeals of delight all round. Philippa, however, suddenly turned serious. 'Does this mean we have to move house again?' She looked dangerously close to tears.
'My goodness no!' answered Jan 'I think we're all happy here aren't we? I certainly don't want to go through that process again.' She looked around for agreement.
Jim answered her with relish 'I think we'll stay here. Maybe we can look at the swimming pool module next?'
The good feeling continued for Jim the following weekend when Tony and Louise took Jim and Jan out for dinner to celebrate Jim's promotion. John and Brenda were also invited along to mark John's semi-retirement.

'Let me propose a toast.' Tony stood up and raised his glass. 'To our new IT and Logistics Director' Jan beamed with pride and Jim looked faintly embarrassed but proud.

Tony continued 'The promotion is not just a thank you for the excellent job that Jim has done, it is also recognition of his newly discovered aptitude for implementing ERP. He will also keep his present position of Special Projects Manager so that he can oversee the next stage of our implementation.'

Jim raised his glass in answer to the smiles and murmurs of congratulation from the people around the table in the smart, fashionable and expensive restaurant. He cleared his throat. 'I always knew that the job in Special Projects was a good one.' He looked meaningfully at Jan and smiled.

Points to reflect on

- Project and change work is ongoing and will be rewarded in the right companies. Embrace the opportunities
- Be strict and formally close the project
- Take the savings to the bank
- At the end – and there needs to be an end – you will have:
✓ added value for customers
✓ added value for shareholders
✓ added value for your people
✓ a mindset to foster continuous improvement

ISBN 1412026563

Made in the USA
Lexington, KY
13 November 2013